Positive Aspects of Animal Welfare

Positive Aspects of Animal Welfare

Special Issue Editor
Silvana Mattiello

MDPI • Basel • Beijing • Wuhan • Barcelona • Belgrade • Manchester • Tokyo • Cluj • Tianjin

Special Issue Editor
Silvana Mattiello
University of Milan
Italy

Editorial Office
MDPI
St. Alban-Anlage 66
4052 Basel, Switzerland

This is a reprint of articles from the Special Issue published online in the open access journal *Animals* (ISSN 2076-2615) (available at: https://www.mdpi.com/journal/animals/special_issues/Positive_Aspects_of_Animal_Welfare).

For citation purposes, cite each article independently as indicated on the article page online and as indicated below:

LastName, A.A.; LastName, B.B.; LastName, C.C. Article Title. *Journal Name* **Year**, *Article Number*, Page Range.

ISBN 978-3-03928-532-7 (Pbk)
ISBN 978-3-03928-533-4 (PDF)

© 2020 by the authors. Articles in this book are Open Access and distributed under the Creative Commons Attribution (CC BY) license, which allows users to download, copy and build upon published articles, as long as the author and publisher are properly credited, which ensures maximum dissemination and a wider impact of our publications.

The book as a whole is distributed by MDPI under the terms and conditions of the Creative Commons license CC BY-NC-ND.

Contents

About the Special Issue Editor vii

Preface to "Positive Aspects of Animal Welfare" ix

Belinda Vigors and Alistair Lawrence
What Are the Positives? Exploring Positive Welfare Indicators in a Qualitative Interview Study with Livestock Farmers
Reprinted from: *Animals* **2019**, *9*, 694, doi:10.3390/ani9090694 1

Monica Battini, Anna Agostini and Silvana Mattiello
Understanding Cows' Emotions on Farm: Are Eye White and Ear Posture Reliable Indicators?
Reprinted from: *Animals* **2019**, *9*, 477, doi:10.3390/ani9080477 31

Marta Brscic, Nina Dam Otten, Barbara Contiero and Marlene Katharina Kirchner
Investigation of a Standardized Qualitative Behaviour Assessment and Exploration of Potential Influencing Factors on the Emotional State of Dairy Calves
Reprinted from: *Animals* **2019**, *9*, 757, doi:10.3390/ani9100757 43

Katrin Spiesberger, Stephanie Lürzel, Martina Patzl, Andreas Futschik and Susanne Waiblinger
The Effects of Play Behavior, Feeding, and Time of Day on Salivary Concentrations of sIgA in Calves
Reprinted from: *Animals* **2019**, *9*, 657, doi:10.3390/ani9090657 55

Sofia Diaz-Lundahl, Selina Hellestveit, Solveig Marie Stubsjøen, Clare J. Phythian, Randi Oppermann Moe and Karianne Muri
Intra- and Inter-Observer Reliability of Qualitative Behaviour Assessments of Housed Sheep in Norway
Reprinted from: *Animals* **2019**, *9*, 569, doi:10.3390/ani9080569 71

Katrin Portele, Katharina Scheck, Susanne Siegmann, Romana Feitsch, Kristina Maschat, Jean-Loup Rault and Irene Camerlink
Sow-Piglet Nose Contacts in Free-Farrowing Pens
Reprinted from: *Animals* **2019**, *9*, 513, doi:10.3390/ani9080513 85

Jaciana Luzia Fermo, Maria Alice Schnaider, Adelaide Hercília Pescatori Silva and Carla Forte Maiolino Molento
Only When It Feels Good: Specific Cat Vocalizations Other Than Meowing
Reprinted from: *Animals* **2019**, *9*, 878, doi:10.3390/ani9110878 97

Alistair B. Lawrence, Belinda Vigors and Peter Sandøe
What Is so Positive about Positive Animal Welfare?—A Critical Review of the Literature
Reprinted from: *Animals* **2019**, *9*, 783, doi:10.3390/ani9100783 103

Silvana Mattiello, Monica Battini, Giuseppe De Rosa, Fabio Napolitano and Cathy Dwyer
How Can We Assess Positive Welfare in Ruminants?
Reprinted from: *Animals* **2019**, *9*, 758, doi:10.3390/ani9100758 123

Dorota Godyń, Jacek Nowicki and Piotr Herbut
Effects of Environmental Enrichment on Pig Welfare—A Review
Reprinted from: *Animals* **2019**, *9*, 383, doi:10.3390/ani9060383 151

About the Special Issue Editor

Silvana Mattiello is Associate Professor of Animal Husbandry at the University of Milan. In 2017, she obtained national qualification as Full Professor. Her main research interest is animal management and welfare in domestic and wild ruminants. She was Principal Investigator of the 7° FP Collaborative Project "Development, Integration and Dissemination of Animal-Based Welfare Indicators, Including Pain, in Commercially Important Husbandry Species, with Special Emphasis on Small Ruminants, Equidae and Turkeys (AWIN)" and coordinated the project "What's in a Meow: Investigating Cat–Human Communication". She is currently coordinating the projects DEMOCAPRA (Development and Dissemination of Innovative Sustainable Farming Systems for Dairy Goats) and VOCAPRA (Automatic Recognition of Dairy Goat Vocalisations). She is presently acting as contact person for the University of Milan in the UFAW/HSA University LINK Scheme. She has published 70 papers in international journals, two books on deer management, and several book chapters on animal behaviour and welfare.

Preface to "Positive Aspects of Animal Welfare"

During recent decades, the interest in animal welfare has been greatly increasing among scientists, veterinarians, farmers, consumers, the general public, and pet owners. As a consequence, several indicators have been developed and used under experimental conditions, at farm level or in the home environment, to assess animal welfare, and specific protocols have been proposed for welfare evaluation of various species in different contexts. Most of the indicators developed so far have focused on negative aspects of animal welfare (e.g., lameness, lesions, diseases, presence of abnormal behaviours, high levels of stress hormones, and many more), and only a few indicators are presently available that highlight the positive aspects of animal welfare. However, the lack of negative welfare conditions represents just a minimum standard that should be guaranteed to animals, and does not necessarily mean that animals are experiencing good welfare conditions nor have a good quality of life. To guarantee high welfare standards, animals should experience positive conditions that allow them to live a life that is really worth living. For these reasons, the new frontiers of animal welfare should aim to reach this high quality of life, and research should now focus on the development and validation of behavioural, physiological, immunological, and productive indicators of positive welfare. Special attention is being paid to animal emotions that can be interpret from facial expressions, body postures, or vocalisations. This Special Issue focuses on the development and validation of indicators of positive welfare or on the refinement of the existing ones, as well as on the identification of suitable living conditions for providing positive welfare to farmed and companion animals.

Silvana Mattiello
Special Issue Editor

Article

What Are the Positives? Exploring Positive Welfare Indicators in a Qualitative Interview Study with Livestock Farmers

Belinda Vigors [1,*] and Alistair Lawrence [1,2]

1 Scotland's Rural College (SRUC), West Mains Road, Edinburgh EH9 3RG, UK; alistair.lawrence@sruc.ac.uk
2 Roslin Institute, University of Edinburgh, Penicuik, EH25 9RG, UK
* Correspondence: belinda.vigors@sruc.ac.uk

Received: 6 August 2019; Accepted: 12 September 2019; Published: 17 September 2019

Simple Summary: Positive animal welfare is a relatively new concept which promotes the welfare benefits of providing animals with greater opportunities for positive experiences, in addition to minimising negative experiences. However, little is known about farmers' attitudes towards this, and their knowledge of positive welfare. This presents a significant hurdle for the promotion of positive welfare indicators on-farm, where their effective implementation may depend on their acceptance by farmers. In response, this study uses qualitative interviews to explore farmers' positive welfare perspectives. It finds that several aspects, reflective of the literature on positive welfare indicators, are evident in farmers' positive welfare-related discussions. These include animal autonomy, play, positive affect, positive human–animal relationships, social interaction and appropriate genetic selection. Such findings provide insights into what farmers consider are relevant to, and indicative of, positive welfare. In addition, this paper explores how farmers see their role in the provision of positive welfare. It finds that farmers largely consider their inputs should be focused on making sure their animals' needs (e.g., resources) are met and negative experiences are reduced, and that positive welfare will arise naturally, or indirectly, out of this. The implications of these findings and their overlaps with the positive welfare literature are discussed.

Abstract: To support the furtherance of positive animal welfare, there is a need to develop meaningful and practical positive welfare indicators for on-farm welfare assessment. Considering the perspectives of farmers is arguably critical in this regard. Doing so helps ensure positive welfare indicators reflect farmers' existing welfare norms and attitudes and, are thus, of practical relevance to them. However, a key issue for such development is the dearth of knowledge on farmers' perspectives of positive welfare. To address this, this study uses qualitative interviews to directly examine livestock farmers' perspectives of positive welfare. Findings reveal that farmers describe elements of positive welfare which are broadly in line with indicators suggested in the positive welfare literature. These elements include animal autonomy, play, positive affect, positive human-animal relationships, social interaction, and appropriate genetic selection. Additionally, this study finds that farmers construct the reduction of negative aspects of welfare as their primary management concern and mostly construct positive welfare as arising indirectly from this. Insights into the importance that farmers of different sectors and systems give to different aspects of positive welfare indicators are also explored. The implications of these findings and the similitudes between farmers' perspectives and the positive welfare literature are discussed.

Keywords: positive animal welfare; positive animal welfare indicators; farmer attitudes; farmer knowledge; qualitative research; free elicitation narrative interviewing

1. Introduction

There is no singular definition of positive animal welfare. However, one broad conception is that it emphasises the welfare relevance of providing animals with opportunities to have positive experiences, in addition to minimising negative experiences [1,2]. As positive animal welfare continues to develop, its furtherance requires suitable indicators for on-farm positive welfare assessment. Here, the primary concern of welfare science will be ensuring such indicators are reliable, valid and feasible [3]. However, research has found that farmers may mistrust the reliability of animal-based welfare indicators [4]. As a result, they may resist changing to systems which increase opportunities for animals to engage in natural behaviours because of economic viability concerns (e.g., cost of investment) [5,6]. This has implications for positive animal welfare, which predominantly focuses on animal-based indicators such as the presence or quality of play behaviours [3,7], social interaction [8] and curiosity driven exploration [9,10]. Engaging with farmers to understand their perspectives on positive welfare and considering this in the development of positive welfare indicators may be critical for their acceptance and implementation at farm level.

Research finds that farmers acceptance and/or implementation of welfare indicators can be influenced by how useful they perceive particular welfare indicators to be, based on their current welfare attitudes and values [4]. In addition, the ability of welfare indicators to function as a decision support tool and be adaptable to local farm conditions can further influence their perceived usefulness [11,12]. Thus, to develop positive welfare indicators which farmers perceive as useful and are therefore willing to accept and implement, it is arguably important to understand their current norms, values, and attitudes surrounding positive animal welfare [11,12]. However, a critical issue in this regard is the dearth of research directly examining farmers' perspectives of positive animal welfare, rather than animal welfare more generally.

In response, this study uses qualitative interviews, informed by the free elicitation narrative interviewing approach [13], to explore farmers' perspectives on positive animal welfare. This paper has three main aims. Firstly, it seeks to uncover whether farmers already engage in practices which (evidence suggests) may generate positive welfare, by exploring how indicators of positive welfare are constructed by farmers within their welfare-related discussions. Secondly, it aims to better understand what farmers consider is relevant to or indicative of positive welfare, by exploring the meaning farmers attach to positive welfare indicators. Thirdly, it seeks to understand whether positive welfare may arise from direct or indirect management practices, by exploring how farmers construct their role in the provision of positive welfare opportunities. Uncovering such insights contributes to the development of positive welfare indicators because it enables an understanding of what farmers think about in relation to positive welfare, and highlights what current practices, attitudes, and norms can be harnessed to effectively introduce positive welfare indicators at the farm level. As Hubbard and Scott [5] note, it is beneficial to involve farmers in the development of scientific techniques, as co-operation between scientists and farmers may support the promotion of higher welfare standards. As positive animal welfare is a newly emerging (and thus not necessarily well known) concept, there is a unique opportunity to explore farmers' related perspectives and behaviours and co-construct positive welfare indicators, from the outset, in line with farmers' language, constructs, values, and norms.

2. Positive Welfare Indicators

Currently, there is no clear or agreed consensus within the literature on what positive welfare should entail or what makes a valid or reliable indicator of positive animal welfare. However, in line with the continued development and growing interest in this concept, several potential indicators for assessing positive welfare in farm animals have been proposed [2,3,14].

One of the most prominent indicators of positive welfare mentioned in the literature is positive affect (see [14–16]). Mellor [10] describes how providing animals with opportunities to engage in goal-directed behaviours which are rewarding can enhance positive welfare, and that such positive affective engagement can, in turn, be used as an indicator of an animal's welfare state. Arguably, other positive welfare indicators suggested within the literature are often those which are believed to contribute to or promote the experience of positive affect, such as social interaction, play and environmental enrichment [14].

Play, for example, is presented in the literature as both an indicator of positive affect [7,14,16] and as a potential indicator of an overall good welfare state [17,18]. Similarly, the nature or quality of social interaction is presented as a potential indicator of positive welfare [3] and linked with positive affect [19]. Here, particular pro-social behaviours such as affiliation (e.g., synchronisation, group bonds) [20], care giving (e.g., maternal behaviour) [21] and social play [17] are thought to be relevant to generating positive animal welfare [19]. Environmental enrichment may also indicate positive welfare [2] as they can stimulate animals to engage in desired goal-directed behaviours [22]. Enrichments can include, for example, mechanical brushes which stimulate self-grooming in dairy cows [23] or the provision of straw which encourage pigs to play or engage in exploratory behaviours [24]. A positive human-animal relationship may also be relevant for positive welfare. For example, Edgar et al. [2] suggest that having positive interactions with stock keepers can support animal confidence. Enabling animals to exert some control over their environment (e.g., have autonomy or agency) has also been linked to positive affect [25] and proposed as a relevant aspect of positive welfare [26]. Closely related to agency, studies have proposed that further indicators of positive welfare could include providing animals with opportunities to exert individual preferences for thermal and physical comfort [2,27]. In addition, Edgar et al. [2] propose the importance of promoting health by "breeding for positive welfare", as indicated by breeding decisions which support the long-term health and welfare of animals, along with a lack of mutilations and effective day-to-day management of health concerns.

Notably, some of this research has involved farmers in the development of positive welfare indicators and frameworks. For example, Edgar et al. [2] first drafted a resource-tier framework of 'good life' opportunities for laying hens through consultation with experts (e.g., welfare scientists and welfare assurance schemes) before interviewing farmers to ascertain their opinions on the framework. More specific to positive welfare, Stokes et al. [28] described using focus groups and consultation interviews to collaborate with dairy and sheep farmers in the development of a framework for assessing positive animal welfare. This included several of the indicators mentioned above, such as positive social interactions and providing animals with some degree of choice within the system (e.g., autonomy). This research interest in working with and considering the farmers' point of view exemplifies the increasing recognition amongst animal scientists that welfare improvement can often hinge on farmer acceptance or buy-in. Nevertheless, to continue the effective development of positive welfare indicators, further research on the perspectives of different farmers is needed, along with a clearer understanding of how they construct and give meaning to specific positive welfare indicators and the underlying factors of influence (e.g., farmers' values, beliefs, attitudes, and social norms).

3. Materials and Methods

This article is based on qualitative interview data collected as part of wider research exploring the perspective of livestock farmers and members of the public on positive animal welfare. This research explored several factors pertaining to societal understanding of positive welfare, including how citizens and farmers framed positive welfare, the aspects of positive welfare evident in their discussions, and the factors underlying and influencing farmers' approach to welfare provision. This present article specifically focuses on the latter two aspects of this interview study.

3.1. Method

The free association narrative interviewing method, designed by Hollway and Jefferson [13], informed the design of the interview protocol. This interview method combines narrative interviewing techniques with the psychoanalytical principle of free association [13]. This means it encourages participants to tell stories and recount personal experiences, within which they have the opportunity to freely elicit factors which they attach meaning to. This ensures participants are unrestricted by any a priori research assumptions which can manifest in narrow or closed questions [13]. This allows the researcher to gain an insight of the participants' perceptions, their representations of the self and the influence of wider social or cultural factors on their opinions [29].

The interview protocol of this study was centred around three focused narrative questions: (i) "Can you tell me about what, in your experience, is a good life for a farm animal?" (ii) "Positive animal welfare—can you tell me about what comes to mind when you hear that?" and; (iii) "Can you describe what motivates you to farm the way you do?" Although the purpose of this research was explained to farmers at the outset, no explanation of what is meant by positive welfare was provided here. The aim was to encourage participants to tell stories (in their words and from their point of view) related to their subjective perceptions of a 'good life' for farm animals and 'positive animal welfare' along with the motivations underlying or driving these perspectives.

Although there is no singular way to conduct a narrative interview [30], the generally recommended approach is to first pose a singular narrative-inducing question (e.g., tell me your experiences of ...) in order to encourage the participant to holistically recount their perspectives on a phenomenon. During this discussion, the researcher does not interrupt the participant or provide any additional prompts but passively takes notes on points of interest freely elicited within the participant's narrative. Once it is clear this main narration has come to a natural end, the researcher can use these notes to ask more probing and detailed questions to gain a deeper understanding of the participant's perspective and experiences [31].

Each of the three focused narrative questions were treated as a separate narrative opportunity and this approach was thus applied to each of the three main questions. Table A1, in the Appendix A, provides an overview of this structure along with some examples of follow-up probing questions, based on the topics mentioned by the participant in response to the initial narrative-inducing question. However, in most interviews the first question resulted in a lengthy response narrative by participants with numerous points raised which required further probing questions. The second question was similar in this regard, while the third question resulted in a shorter, more precise response from participants. This present article is primarily interested in responses to question one and three, as findings from question two have already been analysed in detail in a separate paper (see [32]). However, in instances where responses to question two are relevant to the aims of this paper, they are included in the findings.

3.2. Sample

The population of interest was livestock farmers, from different sectors and with different farming systems (e.g., free-range poultry, zero-grazing dairy etc.). For this reason, the study used a purposive heterogeneous sampling approach to select a diverse range of participants [33,34]. In other words, farmers were not selected on the basis of being a representative sample but one that provided variety in type (e.g., different values, farming systems, etc.), so that as many different insights on positive welfare as possible could be gained.

A variety of means were used to recruit farmers, including advertising on social media, referral from farm advisers and other industry groups, by directly contacting farmers via email or telephone, and attending farmer events (e.g., monitor farm meetings/discussion groups). This resulted in the recruitment of 28 farmers—13 dairy, 10 beef and sheep, two free range egg produces, two mixed farmers (i.e., had pig, poultry, beef and sheep) and one large pig farmer.

The interviews were completed in-person, at the home or farm office of each individual between March and September 2018, in Scotland, UK. On average, interviews lasted 55 min. The majority of participants were male (86%), with a smaller number of female participants (11%) and one participant who preferred not to have their gender recorded. In line with the heterogeneous sampling approach, participants demonstrated a wide variety of different systems within each sector (e.g., organic or non-organic free-range poultry farmers, zero-grazed or pasture-based dairy systems, and indoor or outdoor-wintered beef and sheep). Further demographic information for each participant can be found in Table A2 in Appendix A.

Before the commencement of the interviews, the research was approved by the Scotland's Rural College (SRUC) social science research ethics committee. Participants were given a consent form, detailing the purposes of the research and how their inputs would be used, which they signed. Interviews were also audio-recorded, except for one participant who asked not to be, so detailed notes were taken instead. Interviews were continued until a saturation point, meaning no additional (i.e., 'new') data which had not already been encountered in prior interviews, emerged [35].

3.3. Data Analysis

All of the recorded interviews were transcribed in full (over half by the researcher who conducted the interviews and a smaller number by a professional transcribing service, which were then checked for accuracy by the interviewer) and entered into Max QDA; a software program for assisting qualitative data analysis. The constant comparison method, first described by Glaser and Strauss [36], was used to analyse the data. This involves reading each interview and categorising or 'coding' relevant sections according to the theme of the points raised. For example, codes which emerged in response to the question regarding a 'good life' for farm animals included physical comfort, health, social interaction, and positive affect (amongst several others). Once this initial coding was completed for each interview, they were compared between interviews and related codes and themes were grouped to form overarching superordinate themes. This ensured that the data generated was not influenced by any a priori definitions of positive welfare but were grounded in, or emergent from, the narratives of the participating farmers. These key, overarching themes provided insights into farmers' perspectives on positive welfare and their current practices, as detailed in the subsequent section.

4. Findings

Although farmers were not familiar with the term positive animal welfare, it was evident that several aspects of their current management practices and attitudes towards animal welfare may provide positive welfare opportunities, and a base for their furtherance on-farm. In particular, the positive welfare opportunities of autonomy, play, positive affect, a positive human-animal relationship, social interaction, genetic selection and enrichment emerged as important and recurrent themes within farmers' narratives. These concepts, and the meaning farmers attach to them, are presented in detail in the following sections. The final section describes the perceived role farmers' consider they have in the provision of such welfare factors.

4.1. Autonomy

Autonomy primarily arose within farmers' narratives as providing animals with some degree of choice over being in (i.e., housed) or out (i.e., on pasture).

4.1.1. Choice to Be in or Out

Pasture-based dairy farmers were the most common group to mention providing opportunities for their animals to choose to be in or out. However, the degree to which such autonomy can be provided is influenced by practical constraints, such as the design of the farm, the weather and the land type. The greatest level of free choice was noted by farmers operating a pasture-based robotic system (i.e., Dairy 6 and Dairy 12). Here, the milking herd have constant and complete choice over being inside or out:

> *"Our cows have got free choice. If our cows want to go out, they'll go out, if our cows want to stay in, they stay in".*
>
> (Dairy 12)

Such robotic systems also provide milking cows with free choice over when and how often to be milked.

Those using a more traditional pasture-based system (i.e., grazed during summer, housed during winter, twice-a-day non-robotic milking) note they give their cows choice to be in or out some of the time, often due to the design of the farm:

"The way we can graze one of our fields of paddocks, they can walk into the shed ... Some, if it is a heavy night, they'll go out, they'll eat as much as they can, and then they'll turn around and walk straight back in to the shed and lie down",

(Dairy 3)

or if the weather is bad:

"Maybe it was raining one night, and the cows are not out but we didn't chase them out. We just left them. We said we'll tell you what, if the rain goes off and you want to go out, go out. If you don't want to go out, don't go out. And you see in the morning, all the cows were out. And that was no interference from humans, that was the cow's decision".

(Dairy 5)

In addition, several pasture-based dairy farmers (and most beef and sheep farmers) noted that they looked to their animal's behaviour to decide when to bring them in for the winter, allowing the animals to make that decision:

"And quite often with the Autumn calvers, what we do is we let them out during the day, and once we're finished milking, we'll go and open the gates and they can come back in when they want to, and it is the day they are all standing at the gate waiting for you to open it, that is the day that you know they are not going to be coming out tomorrow. They kind of make that decision for themselves".

(Dairy 1)

Furthermore, most beef and sheep farmers argued that cattle and sheep are best raised outdoors in their 'natural' environment, with several of them arguing for the health and welfare benefits of out-wintering animals. To justify this, they drew heavily from stories of their animals preferring to be outdoors when given the choice, thus highlighting the value they place on animal choice and preference:

"My experience is we have animals that have access to sheds but also have access to woods and even on the stormiest, horrible-ist night .... you'll find the animals lying in the woods".

(Beef and Sheep 6)

The two interviewed egg producers both farmed free range, thus giving their animals full choice over being inside or out. However, both poultry farmers noted that many birds choose not to be outside:

"Chicken by name, chicken by nature. They don't go very far from their shed. They don't, they cannot, they are too frightened to, even if you give them a lot of cover. Remember, all their feed, all their water, everything the hen needs is inside the shed".

(Poultry 2)

The two farmers who kept pigs (Pig 1 and Mixed 2), kept them fully indoors so there was no opportunity for animal autonomy in terms of choosing to be in or out.

4.1.2. Interconnection Between Farmer Values, Farming System and Autonomy

A further interesting finding was an interconnectedness between farmers' personal values, the way in which they farmed, and consequently, how they enabled autonomy. Farmers could be broadly grouped as strongly valuing either a highly structured or a loosely structured farming system. A highly structured system was evident in farmers who emphasised that better welfare was provided by closely controlling, managing and monitoring their animals:

"We've got a lot more control over their diet, over their health and welfare when they're inside"

(Dairy 8)

Farmers with a loosely structured system emphasised that better welfare was provided by limited management intervention and essentially leaving the animal to 'get on with its job':

"If that cow has 14, 15 calves on her own [over her lifetime]; that I never actually have to touch it, apart from tagging the calf, running the bulls and moving the calf from where she calves to the group . . . no intervention at all is ideal . . . she's happy, I'm happy".

(Beef and Sheep 4)

Critically, differences in the way in which autonomy was perceived and enabled were evident between these different structures. Farmers who valued a loosely structured system often argued that limited interference on their behalf meant their animals had greater opportunities to exercise choice and autonomy:

"They spend more time in the field. So they are 20 hours in the field out of 24. And then she is not interacting with anything but her herd mates, because there is no human interaction once you are in the field . . . .they can eat when they want to eat. Walk about when they want to walk about. Drink when they want to drink. Sleep when they want to sleep".

(Dairy 10)

Conversely, animal autonomy in a highly structured system—where farmers valued the welfare benefits gained from having greater control of the animal's environment—was more often discussed in the context of what could be provided within the boundaries of that system. For instance, providing plenty of space in housing so that animals could exert choice on where to lie or move around:

"The bit they're in are 14ft passages. So they've got enough space to run around and do whatever they want. They're not having to barge by each other",

(Dairy 9)

or who to be near:

"So the calves have got a nice big pen to themselves that only they can get in, so it gives them choice so there is not sort of any stress due to there being any forced interactions".

(Beef and Sheep 5)

Critically, issues may arise when the type of farming system does not correspond with the farmer's personal values. This was notable in the narratives of the two poultry farmers. Both noted they farmed free range primarily because this is a publicly desired product. However, both appeared to value more structured and controlled farming systems. Consequently, they often highlighted the welfare issues that could arise within a free range system, due to the lack of close control and monitoring they would prefer:

"There is a big misconception of what free range really is . . . .in the middle of winter when it is terrible weather and they are plodding about in the muck and mud, there is a question of whether it is really higher welfare letting them outside or would it be better sitting them inside".

(Poultry 1)

Indeed, there were evident differences, across all farmers, in how loosely structured and highly structured farmers valued autonomy. The following narrative of a dairy farmer, with a loosely structured system, highlights how highly he values autonomy over other welfare considerations:

"You get things like interaction between cows and dominance . . . .there is all these questions around how that affects happiness and welfare, but I think . . . . having their own choice about what they do overcomes any concern about interaction".

(Dairy 10)

Conversely, the following dairy farmer who values a more structured system demonstrates a cautious approach to autonomy and gives greater importance to close management control:

"I suppose you've got to watch with choice, because cows are like little children, if you give them access to sweeties they'll get a very sore stomach, because they don't stop, they keep going to get more. Aye, I suppose choice to an extent. Just let them go their own way but they still need management".

(Dairy 3)

Overall, autonomy and opportunities for choice was a frequent feature within farmers' discussions of their animal husbandry practices. It was evident from the context of these discussions that farmers are cognisant of their animal's preferences and will often look to 'what would the animals choose to do themselves' when making management decisions. However, there are differences in the extent to which farmers can provide opportunities for autonomy and indeed, individual differences between farmers in the value or relevance placed on providing such opportunities.

4.2. Play

References to play were evident within farmers' narratives. However, the majority of farmers construct play as a non-essential—a behaviour that did not require their direct input—as they perceived animals would engage in it anyway, regardless of what they do:

"So, they [turkeys] like things going on . . . .. just seeing things out and about. They don't physically need to be playing with anything",

(Mixed 2)

and as something which is mostly reserved to young animals:

"Well actually it wears off, you'll see the hogs will skip and jump. The ewes will skip and jump, but you don't see it very often . . . .So it wears off with age. I mean there is no question it is much more obvious in young animals".

(Beef and Sheep 8)

The 'non-essential' status given to play leads to differences in terms of the role farmers perceive they have in promoting play and the relevance given to play as a welfare indicator.

4.2.1. Indirect Promotion of Play

When it comes to promoting play, the majority of farmers perceive their role is minimal as they construct play as an indirect outcome of other factors. Namely: (i) the innate natural behaviours of animals, particularly young animals, as presented above; (ii) good welfare provision (e.g., resource needs are met):

"I would say that [play] is probably only necessary when all of the other needs are met . . . .it is like us with leisure activities, if you have got everything else sorted and then you've got two hours where you have nothing else to do",

(Beef and Sheep 10)

and; (iii) social interaction:

"And that goes again with social interaction, not being on their own. They should be allowed to play, to play with each other".

(Pig 1)

As such, most farmers do not perceive they have to actively promote or directly cause play as it is something which naturally arises from other management-based inputs.

4.2.2. Direct Promotion of Play

However, it was evident that a small number of farmers provide inputs directed towards play:

"We supply toys for the pigs to play with and the pigs will go over and they'll root around the toys ... so they'll have a bit of plastic pipe or something they can chew. Then you'll see them just chewing away with that. Just flicking it around as well".

(Pig 1)

"We sometimes just leave things for them [calves] in the pen to look at, like a tyre ... .and they are all over looking at this tyre thinking what the bloody hell is that. And you take the tyre away and then you maybe leave a bucket ... .and they are playing with this bucket and they think 'oh this is great'. We do that with horses. We buy your dog a toy. Why are calves any different".

(Dairy 5)

Importantly, as noted above, the provision of play objects was almost exclusive to some dairy farmers and the pig farmer. Beef and sheep farmers did not generally provide things for their animals to play with as animals were often noted to naturally make use of objects in their outdoor environment (e.g., branches of trees) or play with one another in their social groups. Interestingly, one farmer noted differences in the play behaviour of lambs when brought indoors to a less 'enriched' environment:

"So when we've ended up with lambs in the shed, finishing lambs in the shed around Christmas time, so they'll probably be seven months old ... and you'd see them playing out in the field but when you've brought them in to the shed, you've got less. They want to play with stuff, they do, and we don't [provide anything]. There is not a lot of interest for them".

(Beef and Sheep 9)

Thus, the direct promotion of play is a management input only a small number of farmers perceive is necessary or worthwhile to engage in.

4.2.3. Play as a Welfare Indicator

Regardless of whether farmers directly promote play or whether it is an indirect outcome of other welfare provisions, it was notable that many farmers use play as a welfare indicator. Specifically, play is seen to indicate that welfare provisions such as social interaction are adequate:

"If social interaction is right, physical comfort is right, you will get play",

(Beef and Sheep 1)

and that overall animal well-being and welfare provision is good:

"I would say that play is the outcome of welfare. I mean can I say, I think it is a correlate, at least, I think it is a strong correlate but I can't say it is the most important thing ... the best I think you could say of play is that it is a correlate of well-being, because I think it is an expression".

(Beef and Sheep 8)

Overall, actively promoting or causing play is not something farmers perceive they are required to do (as animals do it anyway). Rather, the majority of farmers construct play as something which will occur once other welfare needs (e.g., resource provision, social interaction) are met. However, the value farmers give play is more complex than simply saying they do not value it because they do not directly seek to cause it. Instead, they value it as something which is 'nice-to-see' and as a welfare indicator to signal that other welfare requirements of the animal have been met.

4.3. Positive Affect

Positive affect emerged as a strong theme within the findings, with farmers freely eliciting numerous stories of how, when and why their animals experience positive affect. However, farmers' construction of positive affect is complex with several different types being discussed and often used interchangeably.

4.3.1. Contentment

Contentment is, by far, the most frequently mentioned affective state and the one farmers gave the most importance to. To participating farmers, contentment means that an animal is at ease with their environment and that their resource needs are appropriately met. Findings reveal that farmers assigned 'contentment' the role of something they sought to induce or achieve through their management-based inputs and the role of an animal-based output they looked to, to gauge the welfare state of their animals and thus, assess the efficacy of their own management practices:

> *"If I feed my cows before I go for my tea at night, come back out from my tea and they have eaten as much as they possibly can, they are all lying down. So basically if I go back out from my tea and I push the silage in and very few of them would get out of their cubicle and come and eat silage and they are all quite happy and lying down. That is what I am aiming for".*

(Dairy 3)

'Chewing their cud' was, far and wide, the most frequently used phrase amongst farmers of ruminant animals in this context. To say their animals were 'chewing their cud', was to holistically capture and convey all the elements of what contentment meant to them and its role as a welfare indicator. Namely, that their animals 'want for nothing' and are at ease as a result, thus indicating to the farmer they have effectively fulfilled their duty of care to their animals.

However, farmers often used contentment interchangeably with happiness. On numerous occasions, even when directly asked to clarify their meaning, farmers presented 'happy' to mean the same as 'content' and vice versa:

> *"You know cattle and sheep are happy [when] you go into a shed and they are all quiet and they are lying down and sitting chewing their cud, and content. Content is what I would say rather than happy".*

(Beef and Sheep 10)

Overall, farmers highly value 'contentment' as the primary and most important affective state for animals to experience and the affective state they most desire to induce. They rarely distinguish between happiness and contentment. Instead, they appear to assign the same characteristics of animals being at ease, animals' resource requirements being met, and farmer satisfaction (on seeing such contentment) to the context of contentment and happiness. Consequently, it is perhaps more relevant to refer to this desired affective state as happy-content, where the primary indicator of such a positive state to farmers was animals eating, lying down or demonstrating a general state of ease or relaxation:

> *"If it is comfortable and happy and chewing it's cud. Eating plenty and lying down plenty. If it is chilled out and relaxed and healthy, then I'd say it is a happy cow".*

(Dairy 9)

4.3.2. Happy-Energised

Nevertheless, there were some exceptions to farmers construct of 'happy-content' in the presentation of a further positive affective state of 'happy-energised'. Rather than the calm, at ease or passive behaviours of happy-content animals, farmers characterised happy-energised as animals demonstrating much more expressive behaviours such as play, skipping or high-kicking:

"What is a happy cow, well one that does that, runs off down the field with its tail in the air and jumps about. Like it is quite, there are physical behaviours".

(Beef and Sheep 6)

As the above narrative alludes to, discussion of happy-energised indicators were primarily within the context of animals being turned out to pasture (either a new field or being released from winter-housing), interacting with one another or engaging in play behaviours (particularly in young animals). However, unlike happy-content, which farmers actively sought to create, happy-energised was a behaviour which farmers did not perceive required an active role on their part. Rather, they enjoyed seeing it and evidently recognised the enjoyment their animals derived from such experiences but it was not something necessitating managerial input; animals would freely elicit such behaviours themselves.

4.3.3. Pleasure

More broadly, there were numerous instances within farmers' narratives where they described experiences which they considered were pleasurable for their animals. For the most part, this included the apparent pleasure gained from their environment and food and, to a lesser extent, play and autonomy. Pleasure from environmental factors included engaging with environmental enrichments, such as automatic brushes in dairy cows or litter for dust-bathing in poultry:

"That then is them just having a bath and enjoying themselves ... it is good when you do see the birds enjoying the sun and scratching about",

(Poultry 1)

enjoying natural features of their environment, such as good weather and dry pasture:

"So a lot of the time it is just them lying down and enjoying being outside in the sunshine",

(Dairy 6)

or exploring novel environments:

"And we have to laugh at times ... we've got double gates going into an area, and it's only about the size of this kitchen and we've got two pallets of sawdust ... and honest to Christ, see if you leave that gate open for two seconds, the cows are straight in there ... they are really clever".

(Dairy 5)

Farmers also noted the apparent enjoyment their animals experienced from eating particular foods, with some even noting the potential pleasure-related benefits of offering food choices and variety:

"The fact that they clearly do like to eat a lot of things apart from rye grass and clover suggests that they see something in it and I don't know whether it is just simply variety or looking for some mineral nutrient, but it might be nice for them".

(Beef and Sheep 8)

Closely connected to the previous two sections on autonomy and play, it was evident within several farmers' narratives that they perceive their animals gain pleasure from such experiences. Play, from a farmers' point of view, is an evidently pleasurable experience because of the energised and expressive behaviours animals display. Similarly, autonomy is constructed by some farmers to enable positive affect because it allows animals to engage in behaviours they value, such as exploration and social interaction:

> *"Well happy is just having the freedom to know you've got everything that you need and you know you can go and have a wander around and see your surroundings; there's other animals and other fields to see. We turned out 85 calves yesterday and you'd see them skipping and jumping and they stood and grazed for half an hour and then they were skipping and jumping again".*

(Beef and Sheep 3)

Thus, overall, a multifaceted range of affective states are evident within farmers narratives with 'happy-content' emerging as the most highly valued and actively promoted form of positive affect. However, a discussion of affect in this context would not be complete without detailing its strong connection with productivity and animal performance.

4.3.4. Positive Affect and Productivity

From a farmers' point of view, an animal's productiveness is a reflection of its happiness:

> *"I suppose a happy cow has got to be producing milk. If there's something wrong with it, if it's off its milk [yield has reduced], from our dairy point of view we have that as an indicator. If it's lying down, chewing its cud, happy with life, walking evenly on four feet [she is well]".*

(Dairy 9)

As such, for farmers, the productivity output of their animals is objective evidence that their animals are experiencing positive affect. Indeed, the relative value farmers place on 'happy-content' as a welfare indicator appears to arise from its interconnection with animals being stress-free and healthy, both of which are constructed by farmers as primary drivers of productivity. In other words, the promotion of positive 'happy-content' states is partly motivated by a desire to negate negative stresses or health-issues:

> *"For the cow to be contented, it's much better because it's less stressful, therefore she's more content, therefore it's not going to cause any issues with the calf that's inside her, etc.".*

(Beef and Sheep 7)

Nevertheless, regardless of the direct or indirect role of farmers in promoting positive affect, it is evident within farmers' narratives that they are cognisant of animal emotion and that they look to their animal's behavioural and affective expressions as an indicator of how it is experiencing its environment and their management inputs.

4.4. Human-Animal Relationship

The nature and quality of the human-animal relationship—how the farmer interacts with their animals and how animals interact with the farmer—appears to be intrinsically personal to each farmer, with their individual values and life experiences influencing how they navigate this. Nevertheless, the findings reveal some key themes in this regard, consistent across individual farmers and farming sectors.

4.4.1. Calm, Respectful Handling

First and foremost, farmers emphasise the importance of calm, respectful handling and developing familiarity with handlers as a critical husbandry skill; one which positively affects the quality of the human-animal relationship:

> *"Quietly, in a friendly sort of way and for almost all of the year I am able to be respectful and appreciative of them; your patience wears a bit thin over lambing".*

(Beef and Sheep 8)

"Our cows are very quiet, and I would base that on the fact they they've been brought up quietly ... we just work quietly with them and I think we feel that that is just the best way".

(Dairy 11)

This construct was often illustrated further in farmers' discussions of their handling preferences. For instances, some sheep farmers felt limiting or not using dogs to herd sheep enabled more positive human-animal interactions:

"And I think that if you round up sheep with dogs and bring them into the yards and have dogs in the yards keeping them going ... all you're doing is winding them up".

(Beef and Sheep 9)

Interestingly, several beef farmers detailed the use of squeeze crates and curved races, based on the work of Temple Grandin, as a means to reduce stress during handling:

"It's a pneumatic crush so it puts pressure on them. I suppose it's Temple Grandin's theory that if we put a little bit of pressure on them they feel contained ... so they seem to be more comfortable".

(Beef and sheep 3)

"I have the curved race ... stress levels reduce on human and animal hugely".

(Beef and sheep 7)

In addition, dairy farmers who have transitioned to robotic milking detail their experiences of improvements in the quality of the human-animal relationship as a result. Mainly, because they are no longer having to herd or actively move cows (e.g., to collect and bring in for milking), the cows appear more relaxed during human interactions:

"You realise when you move to a robotic system ... how much time you spend chasing cows to the milking parlour ... they get moved a lot. Whereas these ones, the idea is to move them as little as you can, and there is definitely some individuals who have settled on to the robot system ... we had one with a sore foot about a month ago, and you were able to lift it like a horse in the shed, walk up to her, run your hand down the side of her leg and she lifted her foot for you. They are just not expecting you to try and move them somewhere".

(Dairy 6)

Overall, farmers highly value the importance of minimising stress on their animals and this, in part, underlies their motivation to handle their animals in a calm and respectful way.

4.4.2. Knowledge of Animal Characteristics

Knowing the characteristics of individual (if not all) animals within the herd, is emphasised by several farmers as important for the quality of the human-animal relationship, particularly amongst dairy farmers:

"I know every cow out there. If something walks in ... you know straight away if there's something not [right], it sounds silly; there's 300 cows and you can't possibly know them but you do, you spend every day with them".

(Dairy 9)

Indeed, many farmers demonstrate a strong sense of pride in being able to distinguish individual animals within their herd and their unique characteristics. They often see this ability as a critical hallmark of a good stock-person and one which can determine how they specifically interact with individual animals (e.g., more quietly). Thus, possessing knowledge of the habits and personalities of individual animals, where possible, is something several farmers emphasise and give weight to. However, others do note that this becomes more challenging with larger stock numbers and highlight the benefits of technology (e.g., mobility trackers) in this regard.

4.4.3. The Animal-Human Relationship

A further noteworthy facet of the human-animal relationship is the effect of animals on their human carers. There are numerous instances within the narratives of participating farmers detailing how their animals have influenced their human experience on the farm. This was particularly evident in the context of farmers discussing their animals' affective experiences. Namely, when farmers recounted stories about their animals engaging in apparently pleasurable activities and displaying their enjoyment, they nearly always accompanied it with a discussion of the personal enjoyment they experienced from seeing this. In other words, positive affect in animals appears to engender positive affect in farmers, as illustrated in the following narratives:

> *"I like the summer because I move my cattle everyday . . . .I like the physical interaction of the cattle . . . where you go to a group of 120 cattle and those cattle see you arriving and they all get up and wander over to the corner and wait for you to arrive to roll back that reel, and them to walk through . . . .all you are listening to is the rip of that grass . . . and those cattle just as content as they could possibly be . . . .it does make me feel good".*

(Beef and Sheep 6)

> *"There is nothing better than seeing the calves interacting and playing outside. And lambs, you know, when they are a week or two, maybe at the two week old stage, you can't beat watching lambs racing about the field and just being in their natural environment".*

(Dairy 11)

Thus, overall, it is evident that farmers and their animals have an influence on one another, with farmers recognising the impact they have on their animals' daily lives through their handling and management practices, whilst also, often unconsciously, highlighting the impact the behaviour and affective states of their animals has on them.

4.5. Social Interaction

Social interaction is important to farmers. All ubiquitously stressed that farm animals are gregarious and therefore should not be kept alone or prevented from interacting socially. For many farmers, preventing animals from socialising or keeping them alone was a welfare issue:

> *"They get most stressed when they are on their own. That is the thing they dislike more than anything in my view".*

(Beef and sheep 10)

> *"Isolating an animal, . . . if you isolate anybody, human or animal, . . . it's not the best environment for their mental well-being. And pigs are very sociable. I think if you put a pig in a pen on its own it would just, it wouldn't be happy. We would only put them on their own if they are sick".*

(Pig 1)

As such, keeping animals in social groups is considered a basic welfare requirement by all farmers.

4.5.1. Minimising Negative Social Interaction

Farmers primarily consider social interaction as an animal-based output which, similar to play, animals will naturally engage in. Nevertheless, farmers present themselves as having some influence on social interaction, where their management-based inputs can impact the nature of interaction. Here, the majority of farmers construct and confine their management role to the reduction and prevention of negative interactions such as dominance or bullying, usually by separating animals into different groups:

"You would probably separate the older more dominant animals and put them together and let them fight it out and take away the smaller, younger ones, and put them together. But they will still come up with a hierarchy but it might not be quite as dominant and subordinate as it would be if they were all mixed in together",

(Beef and Sheep 10)

designing living spaces so subordinate animals can get away from dominant animals:

"This is one of the benefits of these multi-tier type sheds [as opposed to flat-deck poultry housing] ... the dominant birds tend to take up residence in the top, they will have, you know, their pecking order",

(Poultry 2)

or not unnecessarily mixing animals between groups or herds:

"We have multiple herds, so it is possible for us to change the herds, to mix up the cows in the herd. We try as far as possible not to do that, as you definitely get dominance".

(Dairy 10)

"You don't want to potentially mix pigs, which will create stress".

(Pig 1)

One dairy farmer even noted measuring the impact of such negative social interactions on their cows:

"Some weeks when you put the six or seven new cows in ... .the cell count you'll see will have a little spike and it takes two or three days to come back down again, and that is them just sorting themselves out in the dominance line, or those cows have just come in and are maybe just worked up a little bit".

(Dairy 6)

Thus, it is clearly evident that farmers are cognisant of negative social interactions between animals, are aware that their actions can result in negative social interactions and thus seek to minimise both.

4.5.2. Supporting Social and Maternal Bonds

Beyond the focus on negative social interactions and their preventions, there was some evidence of enabling positive social interactions. Mainly, this centred on the support and maintenance of social bonds. Several dairy farmers, for example, aimed to raise their replacement young-stock (e.g., calves and heifers) in small-groups, enabling them to transition these animals together into the main herd so they could maintain their existing social groups:

"Some of the cows now will stay in batches and you'll see that all the 100s kind of hang about together and the 200s, and they were heifers that were all reared together and they know each other",

(Dairy 12)

or by splitting larger herds into smaller herds:

"I think we will end up with a young cows group ... we will have this as a heifers group and the next shed as our high performing group".

(Dairy 6)

Supporting positive maternal bonds was also particularly important to beef and sheep farmers:

"The parent-offspring bond in animals is pretty strong ... when we have sucklers ... we don't tend to wean until the new calf comes along, so it is sort of natural weaning".

(Mixed 1)

> *"What sheep want is food and they want their lambs and the lambs want their mothers ... I've observed that they want to hang around the animals that they are familiar with and grew up with ... they stick together in the groups that they've known from when they were born or kept together".*

(Beef and Sheep 8)

In particular, they often told stories of specific ewes or specific cows whose high performance they attested to their excellent mothering abilities, such as in the following narrative:

> *"So the ewe that started taking the flock higher [i.e., improved performance] because we used a lot of sons of hers ... .when we shed her lambs to drench them ... ..she is waiting at the end of the race. She's not gone back to the field. She is waiting at the end of the race and she gets one and she waits and gets the other one ... .and we're claiming that she's milkier or has better growth potential ... .but it's her behaviour that has driven that I think. So, I think there is a lot more in the ewes' behaviour than what we're giving credit to".*

(Beef and Sheep 9)

Overall, farmers highly value social interaction and see it as a basic requirement and need of their animals. As such, farmers predominantly view social interaction as a natural behaviour which does not require much management or input from them, beyond intervening to minimise negative social interactions. Nevertheless, they are cognisant of the benefits of positive social interactions to their animals and some seek to support this through the maintenance of social bonds and limited mixing of animals between groups.

4.6. Genetics

The farmers in this study presented genetic selection as a critically important aspect of their management decisions, one which they felt could impact the longer-term welfare and health of their animals. Although the importance placed on genetics was shared across all farmers, it was evident that different farming sectors were motivated by different genetics-related goals.

4.6.1. Welfare Aims Amongst Beef and Sheep Farmers

Beef and sheep farmers, in particular, saw thinking long-term about the type of animals they wanted to breed as a central part of welfare. Three main genetics-related goals emerged within this farming group. First, they emphasised the importance of: (i) selecting for calmness:

> *"So the long term goal of that is to have a flock of ewes here that are happy in their surroundings, in their environment, they're not stressed ... .and an important aspect of that I think is that the ewe's calm enough ... that is a state of mind thing I think for the sheep ... .there is a breeding in the sheep".*

(Beef and Sheep 9)

This genetics-related goal was motivated by wanting to reduce the overall stress their animals experienced while being handled or interacted with. This often led to very specific preventative breeding decisions, such as using polled genetics to avoid the stress caused by de-horning calves:

> *"We've not been a fan of dehorning calves, purely because it's an extra stress on the cow ... so we've been Angus breeders for 20 years, 25 years, so now they're polled, even the Limousine genetics are all polled. So, just one less stress for the cows and the staff as well",*

(Beef and Sheep 3)

and removing 'difficult' animals from the herd to improve its overall 'calmness':

> *"I've got a very active policy here that anything that annoys me at all for any reason is just gone ... .anything that is very flighty, anything that causes me hassle ... which means, and I can see it,*

the herd is getting easier and easier to manage, they're getting quieter . . . everything is calming on its own . . . .I'm not having to go and pull out big calves . . . .so surely, that's what welfare is".

(Beef and sheep 4)

Closely related to this, was beef and sheep farmers' second genetics-related goal of (ii) selecting for ease of management and health. Here, the aim was to produce animals which needed limited intervention (e.g., could calve unassisted):

"We try and use easy-calving strains",

(Beef and Sheep 3)

and had the behavioural and health traits required for the farm's longer-term performance and sustainability.

"Welfare is also an issue if you find that you've got to be intervening all the time using antibiotics, fixing things, bad feet. Surely you should be able to breed all that out so the cow looks after itself, which means that's the best welfare".

(Beef and Sheep 4)

An additional goal of beef and sheep farmers was to (iii) breed to suit the farm system and environment, as opposed to performance-based factors alone. Here, farmers felt that animals would perform better if their breeding suitably equipped them to thrive within the specific topography of the farm or within the farm's management system:

"So the reason we've done well I think is because our Luing cows spend their life here thinking, 'Life's a breeze! We go out in the summer and there is abundant grass and we get pregnant straight away and job's a good'un'. Whereas our Simmental cows, you put them out and they sort of stand at the gate wishing they lived on a better farm, so that is the difference . . . at the moment we've got a drive on to try and change the type of Simmental we have. I think the Simmental has a good role to play [so] we are trying to head for and get a low input Simmental".

(Beef and Sheep 9)

In sum, beef and sheep farmers thought long-term about the type of animal they wanted to breed. They emphasised the importance of selecting for calmness to produce animals that experienced less stress, needed less management intervention and were genetically equipped with the mental or behavioural traits needed to perform well within the specific conditions of their farm.

4.6.2. Welfare Aims Amongst Dairy Farmers

Amongst participating dairy farmers, using genetics to overcome welfare or health issues experienced in the past or to prevent potential future health issues, appeared to be the main goal of their genetics-related decisions. For instance, they used genetics to address lameness and mobility issues:

"We have dealt with lameness by cross-breeding and making sure cows have got black feet",

(Dairy 10)

to support health:

"Even before they are born . . . you are selecting bulls for management traits which are going to lead to cows which are going to last longer, which are going to have fewer problems, . . . keeping that focus on health and welfare right throughout their lives",

(Dairy 1)

and to overcome the 'bull-calf issue' (i.e., lack of a market for bull calves and negative public perception of their culling) by using sexed semen:

"We've gone down the route in the last, just kind of nine months now, of just using sexed semen ... it is nice knowing that in nine months' time, I'm going to get a heifer calf rather than at the end of it you get a bull calf that is, to us, no use really".

(Dairy 9)

Similar to beef and sheep farmers, breeding animals which suit the system of the farm is also an important genetic consideration for dairy farmers:

"The ultimate, this is a question I guess, could the ultimate driver of positive welfare be having the right genetics for your phenotype? And in actual fact I would suspect that is the case, because I need cows that can handle all the rubbish and poor management that I throw at them".

(Dairy 10)

The longevity of their cows also emerged as important. Many farmers noted concerns that productivity-based selection of the past (e.g., higher milk yield) had produced cows which could only do a small number of lactation cycles before health problems occurred (e.g., lameness or mastitis):

"A long life. There are too many animals that go away too early now. That is probably the way we have bred animals though isn't it".

(Dairy 4)

Consequently, several farmers emphasised the importance of milk yield over the animal's life-time (rather than per lactation) and thus selected on factors needed for cows to have a long life (e.g., mobility and health).

"We'd rather reduce the litres and have a cow that lasts longer and gets in calf, rather than have a big showy cow that will only last two or three lactations ... [and that comes] partly from breeding, certainly for more ligament strength .... ..a lot of cows coming on that don't make it that long because udders are hanging off them, dragging on the ground is no use, so there is definitely a bit of breeding in it".

(Dairy 3)

Interestingly, dairy farmers stressed genetics as being a key driver of improved health and welfare in farming in the future, with several commenting on the potential of gene editing in this regard and two farmers noting they genetically test some of their young-stock to inform their breeding decisions. The ubiquitous use of artificial insemination amongst participating dairy farmers may play a role here; having such detail on the genetic characteristics of particular bulls appears to make them more cognisant of the health and welfare implications of genetic selection, when often, trade-offs have to be made and particular genetic traits have to be prioritised:

"We quite often find that we don't end up using the bull with the highest [scores] for milk yield or the highest fat and protein yield because there is one that has got very good feet and legs or one that has got very good udder confirmation or something like that".

(Dairy 1)

Overall, dairy farmers emphasised the importance of using breeding decisions to improve health, minimise previously experienced welfare issues and ensure animals suit the farm environment and can lead a long life. Such detailed discussion demonstrates the value farmers place on genetics, primarily as a means to overcome negative health issues, and in doing so, improve welfare.

4.6.3. Welfare Aims Amongst Pig and Poultry Farmers

The discussion of genetics in terms of health and welfare factors was less notable amongst pig and poultry farmers, potentially as breeding in this sector is generally carried out by specialist breeding companies. However, one poultry farmer, similar to dairy farmers, highlighted the potential for genetics to overcome welfare issues, in this case feather-pecking, but rooted the challenges of this in terms of its effect on productivity:

"We would like to see a genetic change to the hens so eventually they would, we would like to breed this beak [tip] off ... the problem is, you could breed that beak off and lose 50 eggs, no use to me. We're wanting the unachievable perhaps ... unless, you as a consumer are willing to give me £3 a dozen for the eggs. But you're not".

(Poultry 2)

Overall, genetics-related decisions are a central aspect of farmers' management practices and one which they place considerable importance on. Indeed, as presented here, genetics are often used to reduce welfare or health concerns, where the overarching end-goal is to negate issues and improve performance. However, it is notable that many farmers recognise that selecting on productivity-based factors only (e.g., growth or yield) can have adverse health and welfare consequences. Thus, they engage in genetics practices which can indirectly benefit performance and productivity by producing animals which are calmer, are tailored to the specifics of the farm environment, require less intervention, have a longer life, and are, above all, healthy.

4.7. Environmental Enrichment

Overall, farmers place little emphasis or importance on environmental enrichment. This is because, for the most part, farmers perceive that their animals are adequately enriched by the typical environments on a farm. For instance, beef and sheep farmers believe that access to the outdoors provides them with appropriate enrichment, mainly as this enables their animals to interact socially, engage in play behaviours, graze on various plants in hedgerows or in fields, scratch or rub themselves on fence posts or tree branches and explore their fields and surroundings. Consequently, beef and sheep farmers see little requirement for or additional benefits from providing enrichment objects to their animals:

"The animal is much happier outside and we manage our animals in such a way, they get moved every single day 365 days of the year, my cattle are behind an electric fence and that electric fence gets moved every single day ... it is moved every single day, so every single day they are on to a fresh piece of pasture".

(Beef and Sheep 6)

"Sometimes bulls and things will have a bit of a rub on and things like that, so I think it is important that they do have stimuli of some description, even if it is just a branch of a tree ... .I wouldn't say it is cruel to not give them rubs, I would say that is an extra add on rather than an absolute necessity".

(Beef and Sheep 10)

"I think that certainly cattle, as long as they've got interaction with their own, on a day to day basis, they'll not be bored as such, or well ours have got plenty of chance to go and rub and have a scratch and stuff like that. They're quite, my take on cattle is that they are quite routined".

(Beef and Sheep 5)

Nevertheless, dairy farmers, particularly those who have a zero-grazing system, discuss the use of automatic brushes as an enrichment tool for their animals. In most cases, they value the importance of such enrichment objects because they observe and recognise the pleasure and enjoyment their animals get from using them:

"We've got brushes as well, the automatic brushes ... and they absolutely love them. And you can tell if they've been dry [i.e., not milking] and they come in to the parlour again that is probably one of the first things they go and do ... so it must be quite enjoyable for them".

(Dairy 11)

One farmer even noted how observation of cows' use of automatic brushes could highlight other issues:

"The other thing though with brushes, I have a friend who sorted the ventilation in the shed, and after that he said the cows weren't bothering to use the brush and he reckons that because the air is cleaner the cows are staying cleaner, and they are not needing to go and brush themselves as much".

(Dairy 3)

Enrichment, as a management-based input which farmers provide, is most evident in the narratives of pig and poultry farmers. Free range egg producers emphasised the importance of providing litter to hens as an enrichment tool to enable them to dust-bathe:

"The ultimate enrichment for hens is litter .... my hens display all the traits of being perfect, with their feather cover, and they have tremendous litter on the floor. And litter is what it is all about, for a hen, litter is what it is all about, it is like giving that dog a bone. It is just what they want and they display it immediately ... there will not be an inch left where there is not a hen in it. So they must like it".

(Poultry 2)

In addition, enrichment objects were often seen as a way to prevent negative interactions, such as feather-pecking:

"I have big square bales .... ..and they'll jump up and down on that and run around that .... .if they do get bits of feather pecking, I'll go in with maybe cabbages or something. Throw some cabbages down and they can peck away at those so I do have to make sure their environments are rich. I think a bored turkey can be quite a dangerous thing. To each other".

(Mixed 2)

The large commercial pig farmer also discussed the provision of toys to pigs in slatted housing and noted the benefits of such enrichment objects for the pigs:

"I think when toys were first brought in, everyone thought that they were stupid, because they thought what the hell are we doing this for. But you do see them playing with them. And I think that is maybe not a bad thing".

(Pig 1)

Overall, enrichment objects primarily appear within farmers' narratives as something which only animals kept indoors require, those which live most of their life outdoors are perceived as being adequately enriched by being in their 'natural' environment. Consequently, although farmers often discuss and tell stories of the enjoyment and pleasure their animals get when interacting with environmental enrichments, they do not generally place as much importance on this as other factors such as social interaction or autonomy (which are often presented as means of enrichment in and of themselves).

4.8. Positive Animal Welfare from Direct or Indirect Management Practices

A noteworthy finding is how positive welfare opportunities arose from both the direct and indirect management practices of farmers. The context of farmers' animal welfare discussions reveal that there were some welfare indicators they actively seek to provide and there are others which they consider are 'nice-to-see' (i.e., are not essential or high priority). Resource needs, health-related factors and stress reduction often fall into the category of intentional management provision, but animal-based inputs and outputs often fall into a category of 'nice to see'. This is potentially due to the importance farmers place on reducing stress and fulfilling their duty of care by providing an environment which appropriately meets the resource need requirements of their animals (e.g., nutrition, physical comfort, space) and supports their health. Nevertheless, it was evident in farmers' narratives that positive welfare indicators arose from both direct and indirect management practices.

For example, autonomy was directly enabled by farmers who provided their animals with the choice to be in or out, whilst also being indirectly supported by the provision of space and housing design and indirectly affected by the farming system the farmer developed based on their personal values (e.g., loosely structured or highly structured). Similarly, positive social interactions were directly supported by farmers' actions to minimise negative social interactions (e.g., dominance) and support social bonds (e.g., not mixing groups), whilst also indirectly supported by space allocation and housing design. Furthermore, there can be differences in farmers' direct or indirect management practices within a specific welfare indicator, with positive affect being the primary example. Here, all farmers directly sought to induce contentment in their animals, through the provision of desired resource needs. However, more high-arousal emotions, such as pleasure and happy-energised, were indirectly enabled by other management practices such as the provision of enrichment objects or following a change in environment (e.g., turned out to pasture in spring time).

Such insight and knowledge is important because factors which arise from direct inputs are arguably those which farmers consider the most relevant and critical for their animal's welfare. This is pertinent for the development of positive welfare indicators as they reveal what farmers are currently doing, the welfare indicators they are interested in measuring, and what welfare outcomes they particularly value. Farmers may be more likely to accept positive welfare indicators if they are in line with direct practices they already focus on or can arise indirectly from the other management inputs they value.

However, these findings also reveal that farmers do not look at these factors and welfare indicators as singular unrelated entities but see them as an interconnected whole which support and influence one another to impact their animals' well-being. For instance, farmers wrap social interactions up with other factors such as autonomy and play, perceiving that if animals have the freedom to move around and choose who they want to be with, then they can experience positive social interaction, and if they experience social interactions then they are more likely to play. Similarly, if animals are free to socially interact and have all their resource needs appropriately met then they are more likely to experience positive affect. As the below narrative illustrates, farmers' construction of welfare indicators is highly interconnected and are seen as mutually inclusive:

"I think social interaction is very important. I think play comes as part of social interaction, so I don't think play is critical, so you will see that as play being; if social interaction is right, physical comfort is right, you will get play. Which is a good indicator that they are good".

(Beef and Sheep 1)

Thus, the findings of this study reveal that farmers construct such positive welfare indicators to be indirectly supported and enhanced by each other.

A final, noteworthy finding in this regard, is the context-dependency of welfare indicators to farmers. When discussing different aspects and indicators of welfare, many farmers commented on how their specific importance would change depending on the context of the situation. For example, as the farmer below describes, social interaction as a welfare indicator would increase in importance if an animal was kept alone:

"If you are looking at it in the fact that there's an animal on its own, then that becomes number one above anything else; it needs to be with someone else. But if it's looking between a group of ten or a group of 30, it probably doesn't mean a lot. You should never leave an animal on its own".

(Beef and Sheep 9)

Furthermore, as the farmer below explains, the context of perspective, whether it is the animals' or the farmers' perspective, also influences the value placed on particular welfare indicators:

"Well if you give them choice, they wouldn't exercise [but] ... .I think that's crucial for their fitness. But that's me. If the context is what the cow thinks, the cow couldn't give a toss about their fitness".

(Dairy 10)

Such findings reveal the temporality of welfare indicators to farmers. In other words, farmers will often gauge the importance of one welfare indicator against another or against the specifics of the context in which they are in; often, trade-offs have to be made. In addition, farmers continuously stressed the importance of adapting and tailoring their welfare approach and husbandry practices to the specifics of their farm, the type of animals they have (e.g., personality characteristics and breed) and their own personal vision and ethos. Considering all of these factors, it is arguably best to construct farmers' focus on particular welfare indicators as fluid; altering and responding to the needs of different seasons, the specifics of different management systems, adapting to external forces (e.g., market changes) and tailored to the typography and characteristics of their particular farm. In sum, farmers engage in both direct and indirect management practices which may support positive welfare but the nature of that engagement is influenced by the unique characteristics of their personal situation, values and farm.

5. Discussion

To support the development of positive welfare indicators, this paper set out to explore whether and how farmers describe positive welfare indicators within their current practices, the meaning they attach to them and how they construct their role in the provision of positive welfare opportunities. Notably, findings indicate overlaps between the positive welfare literature and farmers' current practices. For example, farmers appear to: value and, to some extent, support positive social relationships [3]; enable a positive human-animal relationship through respectful handling [37]; look to play to assess an animal's welfare state [7]; provide animals with some choice within their environment [25,26,38] and link high-arousal positive affective states (e.g., high-kicking) with situations of contrast (e.g., moving to a new field) [39,40]. However, when looked at as a whole, it is critical to note that the welfare related decisions which farmers gave most importance to are those focused on reducing negative aspects of welfare (e.g., dominance, stress, health issues). Indeed, the central role farmers give to health and stress minimisation in their constructs of good animal welfare and what it entails, may in part explain their focus on health-related genetics decisions and their desire to support contentment in their animals. To see animals content was to indicate the absence of stress and, consequently, signal to farmers the presence of 'good welfare' (i.e., animals resource needs are met, and potential issues have been negated) and 'good health'. Indeed, this ubiquitous desire amongst this study's participants to encourage contentment is perhaps representative of a wider attitude amongst farmers; Spooner et al. [41] also find that pig farmers referred to 'happy' animals when describing a desired condition they sought for their pigs; namely, looking healthy and alert. Thus, it could be argued that there are parallels between farmers' animal welfare-related perspectives and the wider animal welfare literature. Namely, farmers demonstrate a somewhat intuitive awareness of factors relevant for positive welfare, but consider their primary role is to minimise the negative experiences in their animals' lives. Indeed, most farmers appear to hold the view that if they deal with the negative factors impacting their animals, then positive factors will arise naturally, or indirectly, as a result. Therefore, it could be argued that farmers' current practices somewhat overlap with recommendations in the positive welfare literature, but their conscious management decisions are largely based on a belief that welfare is primarily about the reduction of negative issues.

Farmers' perspectives thus appear to possess some similitudes to the traditional animal welfare literature, which focuses more on the reduction of negatives, and the positive welfare literature which promotes the experiences of positives [42]. Determining why such overlaps between science and farmers exist is beyond the scope of this paper, however, it is worthwhile to theorise what may underlie it. One reason for the overlaps between farmers' perspectives and the positive welfare literature may be the animal-based nature of positive welfare indicators. Hubbard and Scott [5], for example, found that when farmers and scientists independently produced welfare assessment measures which focused on using the animal as the source of welfare information, they produced very similar criteria. Thus, it could be theorised that due to their regular interactions with and observations of farm animals, both farmers and scientists form a common understanding of their needs and behaviours and thus consider similar factors as indicators of an animal's welfare state.

Nonetheless, it is important to recognise the limitations of this study and consider the potential for variance or misconceptions in farmers' interpretations of some welfare indicators. For example, behavioural indicators of contentment can easily be confused with apathy and boredom [43]. In addition, the participant sample, although rich in insights and demonstrating a variety of different farming systems (e.g., zero-grazing dairy, out-wintering beef etc.), was mostly made up of dairy and beef and sheep farmers, with a small sample of pig and poultry farmers. Consequently, the sample is primarily composed of participants with an extensive system and a smaller number with an intensive system. As such, further interpretation of and reference to the findings of this study must be cognisant of the context in which they arise and how this affects, and potentially predisposes, positive welfare. For example, beef and sheep farmers, and some dairy farmers (e.g., pasture-based), may have greater opportunities to witness their animals engaging in happy-energised behaviours (e.g., play) because factors which trigger this (e.g., contrast from moving field) are an intrinsic part of their farming system. Similarly, pig and poultry producers may be more likely to use enrichment objects because there are legislative guidelines regarding their provision in these sectors (e.g., [44]). As such, it is important to be cautious when applying the findings of this data; a distinction must be made between farms with extensive and intensive systems, as they may not provide equivalent opportunities for positive welfare. Arguably, a more intensive system may require more direct management interventions to promote positive welfare than an extensive system; further research is needed to better understand the potential distinctions and overlaps between the two in this regard. Nevertheless, beyond these limitations, the insights of this study do highlight several areas which may help support the furtherance and acceptance of positive welfare indicators in livestock farming.

6. Implications for the Development of Positive Welfare Indicators

Findings highlight several key points which may be beneficial to consider when developing positive welfare indicators.

6.1. Farmers' Value System

Previous research has shown that farmers' values can influence their motivation to work with different aspects of animal welfare [45]. This was certainly evident in this study, as farmers who valued a more loosely structured farming system appeared to value animal autonomy more highly, and therefore were more motivated to provide this than those who valued a more highly structured farming system. This has implications for developing and communicating positive welfare indicators, as getting farmer buy-in for their implementation arguably requires indicators to fit with their value orientations. Indeed, the low uptake (10% of holdings in 2011) of the Scottish Government's Animal Welfare Management Programme was thought to be caused, in part, by a lack of fit between individual variations in farming practice and aspects of the scheme [46]. Therefore, developing indicators which can flexibly translate to different farming systems, and thus be communicated in a manner which matches farmers' values and fits with their preferred farming system, may enhance their motivation to employ them and the quality of their implementation.

6.2. The Human-Animal Relationship

An interesting aspect of the human-animal relationship was how farmers' discussions of experiencing personal positive affect (e.g., pride, joy, happiness, pleasure) occurred in the context of their animals experiencing positive affect (e.g., playing, looking content). This indicates the potential for animals experiencing greater opportunities for positive welfare to positively impact the human caretakers. The human-animal relationship literature tends to focus on the impact (often negative) that human caretakers have on farm animals [47,48]. However, there is growing acknowledgment that the human-animal relationship can also affect the on-farm experiences and job satisfaction of the human caretakers [49]. Set within a wider societal context of growing concerns over farmer mental health and well-being in the UK [50,51] and its impact on animal welfare [52], a potentially interesting way to develop and encourage acceptance of positive welfare indicators may be to strengthen awareness of the benefits for the whole farm; the human and the non-human animals. This has previously been noted by Yeates and Main [53], who argued that "where there is a sympathetic human-animal bond, when the human values the good things in an animal's life, enriching the animal's welfare can also enrich the carer's welfare" (p. 294). Thus, emphasising this connection in the development and communication of positive welfare indicators may prove beneficial.

6.3. Indicators Which Signal Farmers are Doing a 'Good Job'

Farmers' discussions of animal welfare, as illustrated in this study, often centre on what they provide and do for their animals, rather than what their animals are experiencing. Consequently, they seek indicators which signal to them that they are doing a good job (i.e., that they are providing what their animals need to thrive and be content). Indeed, although farmers provide a variety of resource, management and animal-based welfare inputs, 'the doing' or provision of good welfare to their animals is what indicates positive welfare to them. Thus, farmers value indicators which inform them of the efficacy of their management practices and inputs. Many of the positive welfare opportunities mentioned throughout could potentially have this role. For instance, play indicates to farmers that their animals' resource or social interaction needs are supported, while contentment indicates they have fulfilled their duty of care. When developing welfare indicators, it is thus worth keeping such findings in mind, as they reveal what farmers want from a welfare indicator and how they wish to use them. To encourage farmer buy-in, it would arguably be of benefit to communicate what information positive welfare indicators could provide to farmers on an animal's welfare state. One particularly important means to do this is to highlight potential links with productivity, as discussed further below.

6.4. Positive Welfare's Connection with Productivity

It is well recognised that farmers almost ubiquitously use the productivity or performance of their animals as an indicator of their welfare state [41,54]. The primary perception of farmers, as discussed in this study, is that animals will not produce if they are not well cared for or not 'happy'. Thus, the motivation of farmers to pursue welfare provisions which support animal productivity could potentially be harnessed to enhance the development and implementation of positive welfare indicators. Research is developing in this regard and studies have found that positive welfare opportunities, such as providing self-grooming brushes to dairy cattle [23] and positive social interactions in pigs [55], can contribute to animal productivity, along with the potential health benefits of positive welfare [42]. Thus, it may be relevant that such links are highlighted in the development of positive welfare indicators and particularly when communicating potential indicators to farmers.

7. Conclusions

This study explored the perspectives of livestock farmers on positive animal welfare to reveal insights relevant to the ongoing development of positive welfare indicators. It finds several positive welfare opportunities are freely elicited by farmers during their discussions of their management practices and their approaches to animal welfare. Namely, some provision for animal autonomy, an awareness of positive affect and a direct desire to promote contentment, a positive human-animal relationship, some support of positive social interactions, the use of genetics to improve health and, in some cases, the provision of environmental enrichments. Such factors fit well with current research on how to provide animals with greater opportunities for positive experiences. However, farmers possess a much more fluid and holistic approach to animal welfare—one that considers the specifics of their animals' current situation, the characteristics of their individual farm, the wider economic environment and one that is underpinned by their personal values and preferences. This presents significant challenges for the development of positive welfare indicators, as there is a need for these to be flexible enough to adapt to the individual characteristics of different farmers and farms. Indeed, as this paper has argued throughout, encouraging the acceptance and implementation of positive welfare at the farm level requires the development of indicators which connect and resonate with farmers' current norms, values and practices. As this study has revealed, there is potential for this by building on existing positive welfare indicators and communicating them in a manner which connects with farmers' value systems, emphasises the whole farm (i.e., human and animal) benefits of a positive human-animal relationship, makes use of indicators which farmers can use to assess the efficacy of their management inputs and highlights connections between positive welfare and productivity.

Author Contributions: A.L. and B.V. conceived and developed the study. B.V. designed the methodology, conducted the interviews, analysed the data and drafted the manuscript. A.L. contributed to the interpretation of the data and the writing and proof-reading of the manuscript.

Funding: This research was supported by funding from the Scottish Government's Rural and Environment Science and Analytical Services Division (RESAS).

Acknowledgments: The authors gratefully acknowledge the input of each of the participants, particularly their generosity in sharing their time and contributing their thoughts and opinions to this study.

Conflicts of Interest: The authors declare no conflicts of interest.

Appendix A

Table A1. Format of narrative interviews with example of follow-up prompt questions.

Narrative-Inducing Question	Main Narration	Questioning Phase (Examples of Follow-up Prompts)
1. Can you tell me about what, in your experience, is a good life for a farm animal?	Active listening—noting further prompting questions	• You mentioned you want them to be 'happy and content'; how would you know that your animals are happy and content? • When you say 'keeping them properly' can you give me some examples of what this involves? • When you say, they are able to 'play with their mates' can you tell me how you manage this? • You mentioned 'giving the cow her choice', can you describe to me how you would do this? • You said that you wanted your animals to be outdoors; from an animal's point of view, what makes this a good life?
2. Positive animal welfare—can you tell me about what comes to mind when you hear that?		• Can you describe what you mean when you say going above and beyond; what does that involve? • You describe it as going beyond just the basics; can you give me some examples of what these may be? • You mentioned 'positive environment'; can you give some examples of what you mean by this? • When you say 'something which is good and kind to the animal' can you give some examples of what being good and kind involves?
3. Can you describe what motivates you to farm the way you do?		• You mentioned seeing things in the past that you didn't like, can you describe to me how this influences how you do things now? • So, what is it about working directly with the animals that you find enjoyable? • Could you perhaps describe some of the specific steps you took in that process, so how you designed your way of farming?

Table A2. Farmers' demographic information.

Sector	Gender	Age	Farm Size (Ha)	Number of Animals	System
Dairy					
1	Male	30–40	130	100–200	Pasture
2	Male	50–60	137	200–300	Pasture
3	Male	18–30	62	100–200	Pasture
4	Male	30–40	343	700–800	Zero-grazed
5	Male	50–60	283	100–200	Pasture
6	Male	30–40	160	300–400	Pasture and robotic milking
7	Male	40–50	344	300–400	Pasture and zero-grazed, non-robotic and robotic milking
8	Male	30–40	100	100–200	Zero-grazed
9	Female	18–30	307	300–400	Zero-grazed
10	Male	40–50	776	1000–1500	Outdoor 365 days/year
11	Female	30–40	687	400–500	Pasture and zero-grazed
12	Male	40–50	176	100–200	Organic and robotic milking
13	Male	40–50	283	800–1000	Zero-grazed
Beef and Sheep					
1	Male	60–70	178	600–700	Indoor-wintered
2	Male	40–50	438	200–300	Indoor-wintered
3	Male	30–40	95	200–300	Outdoor-wintered
4	Male	50–60	230	400–500	Outdoor-wintered
5	Female	30–40	4	<100	Indoor-wintered
6	Male	50–60	100	400–500	Outdoor-wintered
7	Male	40–50	1011	200–300	Indoor-wintered
8	Male	60–70	60	400–500	Outdoor-wintered
9	Male	40–50	500	1000–1500	Outdoor-wintered & Indoor-wintered
10	Prefer not to say	40–50	750	1000–1500	Indoor-wintered
Poultry (laying)					
1	Male	30–40	141	10,000–15,000	Free range and organic
2	Male	50–60	95	120,000–130,000	Free range
Mixed					
1	Male	40–50	54	200–300	Free range (pig and poultry), organic (all species), indoor-wintered (beef), outdoor-wintered (sheep)
2	Male	30–40	230	1000–1500	Free range (poultry), straw-housed (pig), outdoor-wintered (sheep)
Pig					
1	Male	30–40	555	2000–3000	Housed (slats and straw)

References

1. Mellor, D.J. Updating Animal Welfare Thinking: Moving beyond the "Five Freedoms" towards "A Life Worth Living". *Animals* **2016**, *6*, 21. [CrossRef] [PubMed]
2. Edgar, J.L.; Mullan, S.M.; Pritchard, J.C.; McFarlane, U.J.C.; Main, D.C.J. Towards a 'Good Life' for Farm Animals: Development of a Resource Tier Framework to Achieve Positive Welfare for Laying Hens. *Animals* **2013**, *3*, 584–605. [CrossRef]
3. Napolitano, F.; Knierim, U.; Grass, F.; De Rosa, G. Positive indicators of cattle welfare and their applicability to on-farm protocols. *Ital. J. Anim. Sci.* **2009**, *8*, 355–365. [CrossRef]
4. Kirchner, M.K.; Westerath-Niklaus, H.S.; Knierim, U.; Tessitore, E.; Cozzi, G.; Vogl, C.; Winckler, C. Attitudes and expectations of beef farmers in Austria, Germany and Italy towards the Welfare Quality® assessment system. *Livest. Sci.* **2014**, *160*, 102–112. [CrossRef]
5. Hubbard, C.; Scott, K. Do farmers and scientists differ in their understanding and assessment of farm animal welfare? *Anim. Welf.* **2011**, *20*, 79–87.
6. Vanhonacker, F.; Verbeke, W.; Van Poucke, E.; Tuyttens, F.A.M. Do citizens and farmers interpret the concept of farm animal welfare differently? *Livest. Sci.* **2008**, *116*, 126–136. [CrossRef]
7. Ahloy-Dallaire, J.; Espinosa, J.; Mason, G. Play and optimal welfare: Does play indicate the presence of positive affective states? *Behav. Process.* **2018**, *156*, 3–15. [CrossRef] [PubMed]
8. Jensen, M.B. The role of social behavior in cattle welfare. In *Advances in Cattle Welfare*; Elsevier: Amsterdam, The Netherlands, 2018; pp. 123–155. ISBN 978-0-08-100938-3.
9. Mellor, D.J. Positive animal welfare states and encouraging environment-focused and animal-to-animal interactive behaviours. *N. Z. Vet. J.* **2015**, *63*, 9–16. [CrossRef]
10. Mellor, D.J. Enhancing animal welfare by creating opportunities for positive affective engagement. *N. Z. Vet. J.* **2015**, *63*, 3–8. [CrossRef] [PubMed]
11. Rousing, T.; Bonde, M.; Sørensen, J.T. Aggregating Welfare Indicators into an Operational Welfare Assessment System: A Bottom-up Approach. *Acta Agric. Scand. Sect. A Anim. Sci.* **2001**, *51*, 53–57.
12. Vaarst, M. Evaluating a concept for an animal welfare assessment system providing decision support using qualitative interviews. *Anim. Welf.* **2003**, *12*, 541–546.
13. Hollway, W.; Jefferson, T. The free association narrative interview method. In *The SAGE Encyclopaedia of Qualitative Research Methods*; Given, L., Ed.; Sage: Sevenoaks, CA, USA, 2008; pp. 296–315.
14. Boissy, A.; Manteuffel, G.; Jensen, M.B.; Moe, R.O.; Spruijt, B.; Keeling, L.J.; Winckler, C.; Forkman, B.; Dimitrov, I.; Langbein, J.; et al. Assessment of positive emotions in animals to improve their welfare. *Physiol. Behav.* **2007**, *92*, 375–397. [CrossRef]
15. Mellor, D. Animal emotions, behaviour and the promotion of positive welfare states. *N. Z. Vet. J.* **2012**, *60*, 1–8. [CrossRef]
16. Marcet Rius, M.; Cozzi, A.; Bienboire-Frosini, C.; Teruel, E.; Chabaud, C.; Monneret, P.; Leclercq, J.; Lafont-Lecuelle, C.; Pageat, P. Selection of putative indicators of positive emotions triggered by object and social play in mini-pigs. *Appl. Anim. Behav. Sci.* **2018**, *202*, 13–19. [CrossRef]
17. Held, S.D.E.; Špinka, M. Animal play and animal welfare. *Anim. Behav.* **2011**, *81*, 891–899. [CrossRef]
18. Lawrence, A. Consumer demand theory and the assessment of animal welfare. *Anim. Behav.* **1987**, *35*, 293–295. [CrossRef]
19. Rault, J.-L. Be kind to others: Prosocial behaviours and their implications for animal welfare. *Appl. Anim. Behav. Sci.* **2019**, *210*, 113–123. [CrossRef]
20. Boissy, A.; Nowak, R.; Orgeur, P.; Veissier, I. Social relationships in domestic ruminants: Constraints and means for the integration of the animal into its environment. *Prod. Anim. Paris Inst. Natl. Rech. Agron.* **2001**, *14*, 79–90.
21. Newberry, R.; Swanson, J. Implications of breaking mother—Young social bonds. *Appl. Anim. Behav. Sci.* **2008**, *110*, 3–23. [CrossRef]
22. Mellor, D.J. Positive animal welfare states and reference standards for welfare assessment. *N. Z. Vet. J.* **2015**, *63*, 17–23. [CrossRef]
23. Keeling, L.J.; De Oliveira, D.; Rustas, B.O. *Use of Mechanical Rotating Brushes in Dairy Cows—A Potential Proxy for Performance and Welfare*; Wageningen Academic Publishers: Leeuwarden, The Netherlands, 2016.

24. Marcet-Rius, M.; Kalonji, G.; Cozzi, A.; Bienboire-Frosini, C.; Monneret, P.; Kowalczyk, I.; Teruel, E.; Codecasa, E.; Pageat, P. Effects of straw provision, as environmental enrichment, on behavioural indicators of welfare and emotions in pigs reared in an experimental system. *Livest. Sci.* **2019**, *221*, 89–94. [CrossRef]
25. Špinka, M.; Wemelsfelder, F. *Environmental Challenge and Animal Agency*; CAB International: Wallingford, UK, 2011; pp. 27–44.
26. Špinka, M. Animal agency, animal awareness and animal welfare. *Univ. Fed. Anim. Welf.* **2019**, *28*, 11–20. [CrossRef]
27. Mullan, S.; Edwards, S.; Butterworth, A.; Whay, H.; Main, D. A pilot investigation of possible positive system descriptors in finishing pigs. *Anim. Welf.* **2011**, *20*, 439–449.
28. Stokes, J.E.; Main, D.C.J.; Mullan, S.; Haskell, M.J.; Wemelsfelder, F.; Dwyer, C.M. Collaborative Development of Positive Welfare Indicators with Dairy Cattle and Sheep Farmers. In Proceedings of the Measuring Animal Welfare and Applying Scientific Advances—Why is it Still so difficult? Royal Holloway, University of London, Surrey, UK, 27–29 June 2017; p. 133.
29. Squire, C. *Approaches to Narrative Research*; ESRC National Centre for Research Methods: London, UK, 2008; pp. 1–60.
30. Squire, C. Experience-centred and culturally-oriented approaches to narrative. In *Doing Narrative Research*; Andrews, M., Squire, C., Tamboukou, M., Eds.; SAGE Publications: London, UK, 2008; pp. 41–64.
31. Jovchelovitch, S.; Bauer, M.W. Narrative interviewing. In *Qualitative Research with Text, Image and Sound: A Practical Handbook for Social Research*; Bauer, M.W., Gaskell, G., Eds.; SAGE Publications: London, UK, 2000; pp. 57–74.
32. Vigors, B. Citizens' and Farmers' Framing of 'Positive Animal Welfare' and the Implications for Framing Positive Welfare in Communication. *Animals* **2019**, *9*, 147. [CrossRef]
33. Robinson, O.C. Sampling in Interview-Based Qualitative Research: A Theoretical and Practical Guide. *Qual. Res. Psychol.* **2014**, *11*, 25–41. [CrossRef]
34. Palinkas, L.A.; Horwitz, S.M.; Green, C.A.; Wisdom, J.P.; Duan, N.; Hoagwood, K. Purposeful sampling for qualitative data collection and analysis in mixed method implementation research. *Adm. Policy Ment. Health* **2015**, *42*, 533–544. [CrossRef]
35. Saunders, B.; Sim, J.; Kingstone, T.; Baker, S.; Waterfield, J.; Bartlam, B.; Burroughs, H.; Jinks, C. Saturation in qualitative research: Exploring its conceptualization and operationalization. *Qual. Quant.* **2018**, *52*, 1893–1907. [CrossRef]
36. Glaser, B.G.; Strauss, A.L. *The Discovery of Grounded Theory: Strategies for Qualitative Research*; Aldine: Chicago, IL, USA, 1967.
37. Ceballos, M.C.; Sant'Anna, A.C.; Boivin, X.; Costa, F.D.O.; Carvalhal, M.V.D.L.; Paranhos da Costa, M.J.R. Impact of good practices of handling training on beef cattle welfare and stockpeople attitudes and behaviors. *Livest. Sci.* **2018**, *216*, 24–31. [CrossRef]
38. Nicol, C.J.; Caplen, G.; Edgar, J.; Browne, W.J. Associations between welfare indicators and environmental choice in laying hens. *Anim. Behav.* **2009**, *78*, 413–424. [CrossRef]
39. Rushen, J.; Butterworth, A.; Swanson, J.C. Animal Behavior and Well-Being Symposium: Farm animal welfare assurance: Science and application. *J. Anim. Sci.* **2011**, *89*, 1219–1228. [CrossRef]
40. Loranca, A.; Torrero, C.; Salas, M. Development of play behavior in neonatally undernourished rats. *Physiol. Behav.* **1999**, *66*, 3–10. [CrossRef]
41. Spooner, J.M.; Schuppli, C.A.; Fraser, D. Attitudes of Canadian Pig Producers toward Animal Welfare. *J. Agric. Environ. Ethics* **2014**, *27*, 569–589. [CrossRef]
42. Lawrence, A.B.; Newberry, R.C.; Špinka, M. Positive welfare: What does it add to the debate over pig welfare? In *Advances in Pig Welfare*; Špinka, M., Ed.; Herd and Flock Welfare; Woodhead Publishing: Duxford, UK, 2018; pp. 415–444. ISBN 978-0-08-101012-9.
43. Wemelsfelder, F. How Animals Communicate Quality of Life: The Qualitative Assessment of Behaviour. *Anim. Welf.* **2007**, *16*, 1–12.
44. Scottish Government. *Pigs: Codes of Practice for the Welfare of Pigs*; Scottish Government: Edinburgh, Scotland, 2012; ISBN 978-1-78045-754-3.
45. Hansson, H.; Lagerkvist, C.J.; Vesala, K.M. Impact of personal values and personality on motivational factors for farmers to work with farm animal welfare: A case of Swedish dairy farmers. *Anim. Welf.* **2018**, *27*, 133–145. [CrossRef]

46. FAWC. *Farm Animal Welfare: Health and Disease*; Farm Animal Welfare Committee: London, UK, 2012; pp. 1–97.
47. Hemsworth, P.H. Human–animal interactions in livestock production. *Appl. Anim. Behav. Sci.* **2003**, *81*, 185–198. [CrossRef]
48. Devitt, C.; Kelly, P.; Blake, M.; Hanlon, A.; More, S.J. An Investigation into the Human Element of On-farm Animal Welfare Incidents in Ireland. *Sociol. Rural.* **2015**, *55*, 400–416. [CrossRef]
49. Edwards-Callaway, L.N. 4. Human–animal interactions: Effects, challenges, and progress. In *Advances in Cattle Welfare*; Tucker, C.B., Ed.; Woodhead Publishing: Duxford, UK, 2018; pp. 71–92. ISBN 978-0-08-100938-3.
50. Barlow, G. Farmers on the Edge. Available online: https://www.bbc.com/news/av/uk-47888402/farmers-struggling-with-mental-health (accessed on 20 May 2019).
51. Howley, P.; Dillon, E.; Heanue, K.; Meredith, D. Worth the Risk? The Behavioural Path to Well-Being. *J. Agric. Econ.* **2017**, *68*, 534–552. [CrossRef]
52. FAWC. *Opinion on the Links between the Health and Wellbeing of Farmers and Farm Animal Welfare*; Farm Animal Welfare Committee: London, UK, 2016; p. 31.
53. Yeates, J.W.; Main, D.C.J. Assessment of positive welfare: A review. *Vet. J.* **2008**, *175*, 293–300. [CrossRef]
54. Skarstad, G.A.; Terragni, L.; Torjusen, H. Animal welfare according to Norwegian consumers and producers: Definitions and implications. *Int. J. Sociol. Food Agric.* **2007**, *15*, 74–90.
55. Camerlink, I.; Bijma, P.; Kemp, B.; Bolhuis, J.E. Relationship between growth rate and oral manipulation, social nosing, and aggression in finishing pigs. *Appl. Anim. Behav. Sci.* **2012**, *142*, 11–17. [CrossRef]

© 2019 by the authors. Licensee MDPI, Basel, Switzerland. This article is an open access article distributed under the terms and conditions of the Creative Commons Attribution (CC BY) license (http://creativecommons.org/licenses/by/4.0/).

Article

Understanding Cows' Emotions on Farm: Are Eye White and Ear Posture Reliable Indicators?

Monica Battini *, Anna Agostini and Silvana Mattiello

Dipartimento di Medicina Veterinaria, Università degli Studi di Milano, 20133 Milan, Italy
* Correspondence: monica.battini@unimi.it

Received: 26 June 2019; Accepted: 22 July 2019; Published: 24 July 2019

Simple Summary: It is globally recognized that emotions are important elements of farm animals' life. However, scientific understanding regarding how to measure and interpret positive emotional states is currently lacking. This study investigated whether eye white and ear posture can reliably help in the interpretation of mood and level of excitement in dairy cows. We found that eye white and ear posture are strongly correlated, and that can be used as complementary measures to interpret emotions. Daily access to pasture has beneficial effects on cows' emotions. Animals are more relaxed than in any other context, with most of the animals exhibiting half-closed eyes and ears hung down or backwards. The cows were found to be particularly excited during the execution of a human-animal relationship test, showing eye white clearly visible and ears directed forwards, towards the assessor. Housing has an important effect on cows' emotions: the lower the competition for resources (i.e., in case of more feeding places or cubicles than the number of animals), the lower the level of excitement. This research is a further step towards the use of indicators able to measure emotions in dairy cows and can contribute to enhance animals' quality of life on farm.

Abstract: Understanding the emotions of dairy cows is primarily important in enhancing the level of welfare and provide a better life on farm. This study explored whether eye white and ear posture can reliably contribute to interpret valence and arousal of emotions in dairy cows. The research was conducted in five Italian dairy farms. Four hundred and thirty-six photographs of cows' heads were scored (four-level), according to the eye white and ear posture during feeding, resting, pasture, and an avoidance distance test at the feeding rack (ADF test). Eye white and ear posture were significantly correlated and influenced by the context ($P = 0.001$). Pasture was the most relaxing context for cows (67.8% of half-closed eyes; 77.3% ears hung down or backwards). The excitement during ADF test was high, with 44.8% of eye white being clearly visible and ears directed forwards to the approaching assessor (95.5%). Housing and management mostly influenced emotions during feeding and resting ($P = 0.002$ and $P = 0.001$, respectively): where competition for feeding places and cubicles was low, the cows showed the highest percentages of half-closed eyes and ears backwards or hung down. This research supports the use of eye white and ear posture as reliable indicators of emotions in dairy cows.

Keywords: positive indicators; emotions; valence; arousal; dairy cows; eye white; ear posture

1. Introduction

The use of animal-based indicators to assess the welfare of farm animals is broadly accepted, as they are more indicative than resource-based measures of the actual animal experience [1]. So far, the protocols used to assess the welfare of dairy cows mostly include animal-based indicators that evaluate physical conditions (e.g., body condition score, presence of injuries, lameness) or behaviors (e.g., agonistic interactions, social behavior). Only one indicator, namely the Qualitative Behavior Assessment (QBA), is commonly adopted to evaluate the "Positive emotional state". QBA relies on the ability of observers to judge and integrate perceived details of animals' body language and posture into descriptors of low/high arousal and positive/negative valence. Studies that were conducted in different species and contexts (e.g., [2–4]) showed that this method is valid and reliable. However, the need for extensive training of the observers and the difficulties in validating QBA in on-farm conditions may impair its use [5,6].

It is now globally recognized by researchers that emotions are part of the complex life of dairy cows [7], and understanding how animals communicate their emotional state or cope with the environment is important in ensuring a better quality of life and high levels of welfare on farm [8]. Consequently, the most recent trends in animal welfare research are now focusing on the investigation of valid and feasible indicators that are able to measure valence and arousal of emotions in dairy cows [5]. According to dimensional theories [9], emotions can be described as moving in a *continuum* along two axes: valence, which expresses the mood (positive or negative), and arousal, which defines the level of excitement (low or high). Obviously, valence and arousal are strictly linked, and possible interactions need to be considered [10]. Negative emotions often coincide with high arousal (e.g., separation from a group [10]), but high arousal can also be found during positive situations (e.g., receiving highly palatable food [11]), and positive emotions are often characterized by low arousal (e.g., grooming [7]).

To date, many studies have focused on eye white and ear posture as potential promising indicators for interpreting emotions in dairy cows. The majority of research that has been performed to explore the reliability of eye white as an indicator of emotional state was performed on cows (e.g., [7,11–16]), but few studies have also been conducted on sheep [10,17–19]. Research conducted so far confirms the ability of this indicator to capture the variations of arousal of animals that were subjected to different stimuli: eye white percentage or eye aperture significantly decrease when cows and sheep experience a low arousal event, usually with a positive valence, such as gentle stroking [7,10]; conversely, they increase during high arousal situations with either negative (suddenly opened umbrella [13], denied access to visible food [14], calf-cow separation [15], separation from the group [10], being fed with inedible woodchip [11,20]) or positive valence (highly desirable concentrate feed [11]). Only one study does not report differences in the percentage of visible eye white depending on the context (feeding *versus* claw trimming in dairy cows) [21]. However, the authors found a possible confounding breed effect that may have masked the potential treatment effect: this might be due to differences in eye coloration patterns, with Red Holstein cows having more contrast between eye white and iris and Brown Swiss having less contrast due to their darker eye white [21].

Animals use a complex set of body postures and facial expressions to communicate their emotions. In particular, ear postures are largely adopted by a wide range of farm animals in social communication and to express internal states [22]. Ear postures can reveal different emotions, depending on the species; hence, specific studies are needed to gather species-specific information from this indicator [23]. Ruminants have highly developed muscles around their ears, which enables them to independently move ears in many different ways [17]. Therefore, the study of ear posture as an indicator of emotions seems particularly promising. Studies on ear posture have been conducted in dairy cows, as well as in other ruminants (sheep [17,24–26], goats [27]). Backwards or hanging ears in dairy cows and sheep are associated with the positive emotional states of low arousal, as might be induced by stroking or grooming [23,28,29]; consistently, goats spend more time with ears in a forward position when they observe pictures of other goats' faces that were taken in negative situations [27].

This work aimed to investigate whether the visible eye white and ear posture can reliably contribute to interpret the valence and the arousal of emotions in dairy cows. First, we investigated the relationship between eye white and ear posture. Secondly, each indicator was evaluated in different contexts (feeding, resting, pasture, and avoidance distance test at the feeding rack) that are supposed to elicit different emotions. Finally, within the same context, we investigated the effect of different housing and management conditions on the emotional state of dairy cows.

2. Materials and Methods

The study was conducted from March to June 2018 in five dairy farms that were located in Northern Italy. Table 1 presents the main farm characteristics.

Table 1. Main characteristics of dairy farms involved in the study.

Farm ID	Number of Lactating Cows	Breeds	Husbandry System	Cubicle: Cow Ratio	Feeding Place: Cow Ratio	Feeding	Access to Pasture
1	112	Holstein	Loose house, cubicles	0.89	0.69	Total mixed ration	No
2	120	Holstein	Loose house, cubicles	1.08	0.68	Total mixed ration	No
3	49	Holstein	Tie stall	1.00	1.00	Total mixed ration	No
4	52	Holstein	Loose house, cubicles	1.15	1.00	Total mixed ration	No
5	50	80% Pezzata Rossa Italiana, 20% Holstein	Loose house, cubicles	1.12	1.72	Ventilated hay and fresh grass	Yes

More than 500 photographs of cow's heads were taken by the same assessor from one random side of each cow in four different contexts: (1) feeding (head in the feeding rack), (2) resting (cows lying down, either sleeping or ruminating), (3) pasture (cows engaged in different activities while at pasture, such as grazing or lying; this context could be observed only in farm 5, where pasture was available), and (4) an avoidance distance test at the feeding rack (ADF test). The test was performed by a person that was unknown to the animals, who stood still at 200 cm distance in front of each cow at the feeding rack (not restrained) and slowly approached the animal with the arm lifted (45°) and the hand palm directed downwards after having established a reciprocal visual contact, until the first avoidance reaction of cows [30]. ADF test could be performed only in farm 1, thanks to the simultaneous presence of two assessors: one performing the test, the other one taking photographs. Resting was never observed in Farm 2, due to the limited observation time.

The photographs were taken from a distance while using a Canon 650d, mounting a Canon EF 70–300 mm f/4-5.6 IS USM II telephoto lens in order to minimize the assessor's effect on the animals. The photographs were scored from 1 to 4 according to eye white and ear posture, as described in Figure 1; Figure 2. For eye white, we proposed this four-level classification based on eye aperture and if the white was visible or not, instead of the computerized measurement that was proposed by Sandem et al., 2002 [7], in order to improve the feasibility of this indicator. The classification adopted for ear posture is the same as described by [23]. For both of indicators, the lower scores correspond to the highest level of arousal/excitement, and vice versa.

A non-parametric Spearman rank correlation test was used to measure the degree of association between the eye white and ear posture scores. A Chi Square test was used for testing the differences in the proportion of eye white and ear posture classes among contexts, and among farms within each context (only feeding and resting contexts were considered for this last analysis, as the pasture and ADF test could only be recorded in farms 5 and 1, respectively).

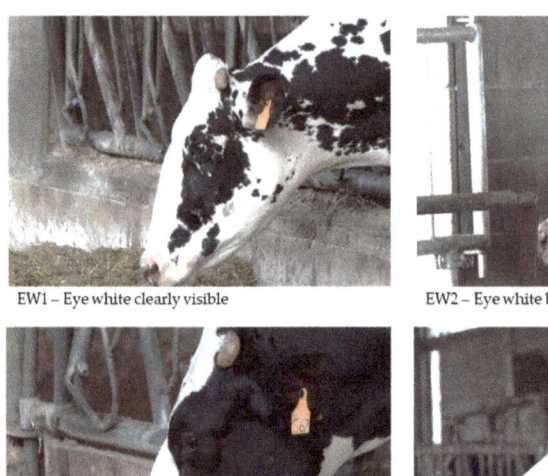
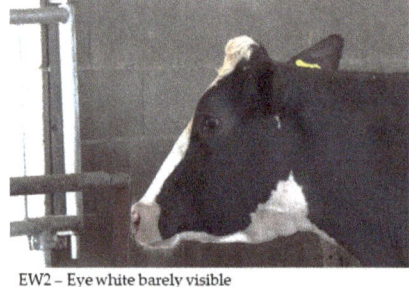

Figure 1. Eye white classification.

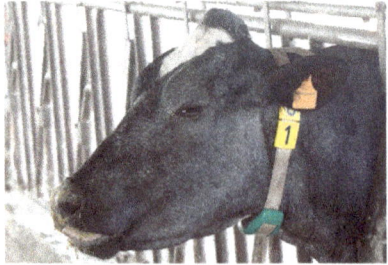
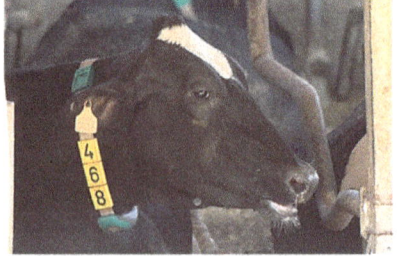

Figure 2. Ear posture classification (previously defined by Proctor and Carder, 2014 [23]).

3. Results

Some of the photographs of cows' heads had to be discarded due to the incomplete visibility of eyes or ears. A total of 436 photographs were retained for eye white classification (feeding = 129; resting = 181; pasture = 59; ADF test = 67) and 489 photographs were used for ear posture classification (feeding = 137; resting = 219; pasture = 66; ADF = 67). Only 429 photographs allowed for the analysis of correlation between eye white and ear posture, as both the eye and ear were clearly visible.

Eye white and ear posture were clearly associated: a high portion of visible white corresponded to ears held upright or directed forwards, whereas half-closed eyes corresponded to ears that were held backwards or loosely hung down (Figure 3); the correlation between eye white and ear posture scores was statistically significant (Spearman rank correlation test: $\rho = 0.570$; $P = 0.001$).

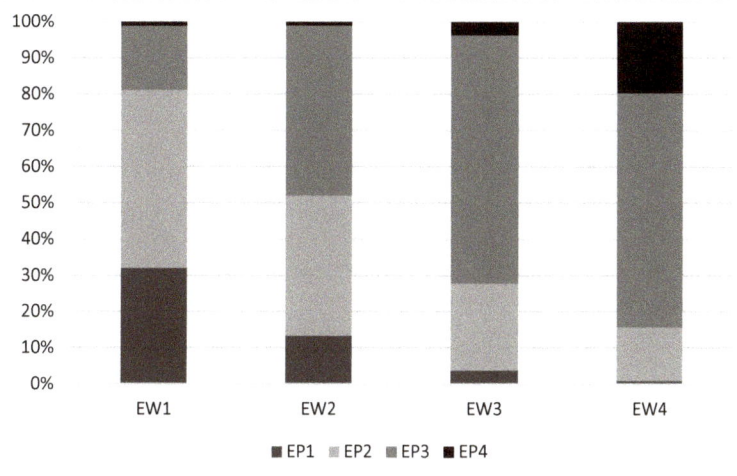

Figure 3. Association between eye white and ear posture: the highest percentage of eye white clearly visible corresponds to the highest percentage of ear in upright position, and to the lowest percentage of ear hung down loosely.

The visible eye white and ear posture were both significantly influenced by context ($P = 0.001$; Figures 4 and 5).

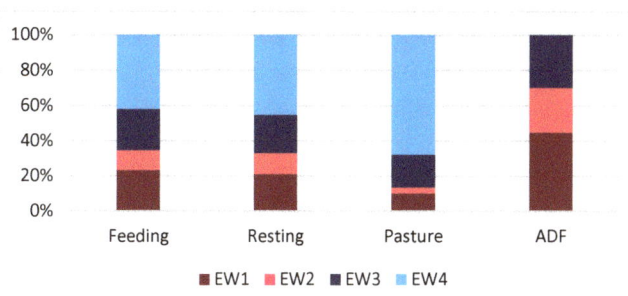

Figure 4. Percentage distribution of the four eye white classes in each context (feeding, resting, pasture, ADF–avoidance distance test at the feeding rack).

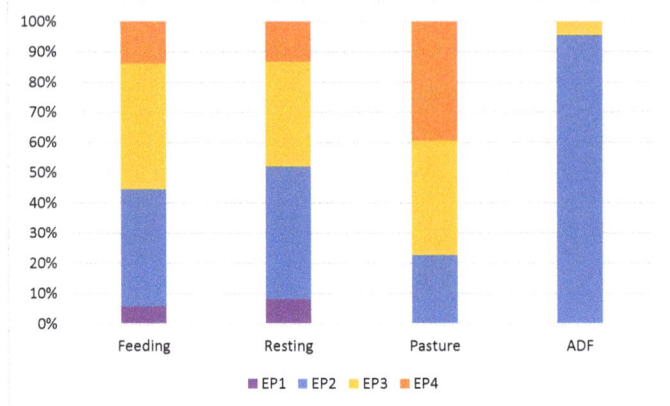

Figure 5. Percentage distribution of the four ear posture classes in each context (feeding, resting, pasture, ADF–avoidance distance test at the feeding rack).

In particular, the highest percentage of half-closed eye (EW4) was associated to the low arousal and positive experience of pasture (67.8%), followed by resting (45.3%) and feeding (41.9%). Half-closed eye was never observed during the ADF test. During the ADF test, the highest percentage of eye white clearly visible (EW1 = 44.8%) was recorded; the highest percentage of cows with normally open eye (EW3) was recorded during the ADF test (29.8%).

During pasture, most of the cows loosely held the ears down (EP4 = 39.4%) or backwards (EP3 = 37.9%); most of the cows held the ears backwards (EP3 = 41.6%) or with the ear pinna directed forwards (EP2 = 38.7%) during feeding. During ADF, nearly the totality of cows directed the ear pinna forwards (EP2 = 95.5%). The cows only showed ear upright (EP1) during feeding (5.8%) and resting (8.2%).

The more common contexts, i.e., feeding and resting, were further analyzed in order to highlight the differences among farms. For eye white, statistical differences among the farms were observed for both feeding ($P = 0.002$) and resting ($P = 0.001$).

Most of the cows in farm 4 showed half-closed eye during feeding (EW4 = 81.8%); the cows in farm 5 showed eye from half-closed to normally open (EW4 = 50%; EW3 = 31.2%). The cows in farm 3 showed eye white ranging from barely to clearly visible (37.5% for both EW2 and EW1) (Figure 6a).

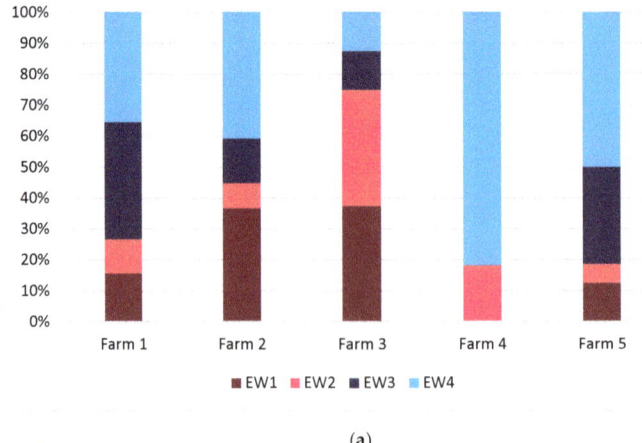

(a)

Figure 6. *Cont.*

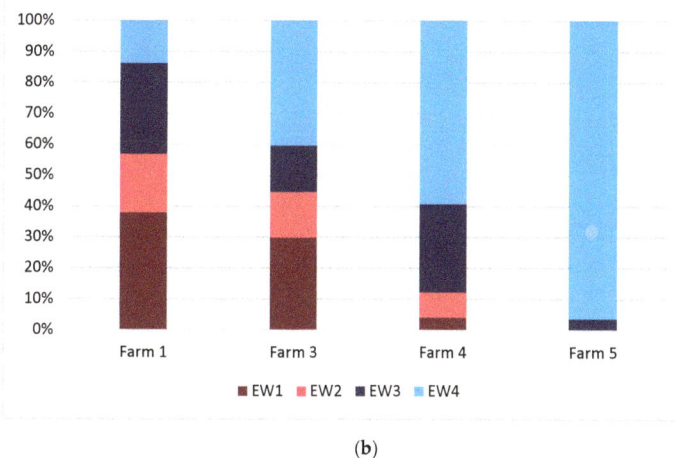

(b)

Figure 6. Results of the eye white in the farms included in the study. (**a**) during feeding; and, (**b**) during resting.

During resting, nearly the totality of cows showed half-closed eye during resting in farm 5 (EW4 = 96.3%), whereas the highest percentages of eye white clearly visible (EW1) was recorded in Farm 1 (37.9%) and 3 (29.8%) (Figure 6b).

As to ear posture, no statistical difference was found among farms during feeding, as in all the farms most of the cows generally held the ear backwards (EP3) and forwards (EP2) (Figure 7a). However, it is worth noticing that the highest percentage of cows with ears held backwards was recorded in farm 3 (EP3 = 50%) and the highest percentage of cows with ears held forwards was found in farm 4 (EP2 = 54.5%). The cows rarely held the ear in upright position during feeding, except in farm 3 (12.5%; Figure 7a).

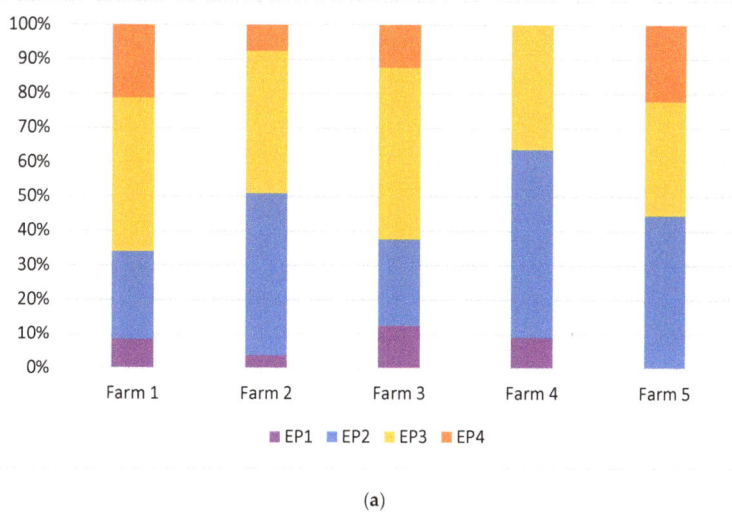

(a)

Figure 7. *Cont.*

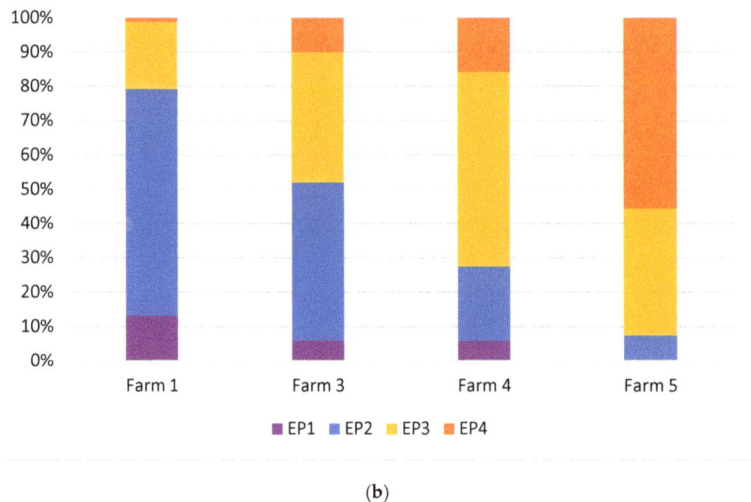

(b)

Figure 7. Results of the ear posture in the farms included in the study. (**a**) during feeding; (**b**) during resting.

Ear posture statistically differed among farms during resting ($P = 0.001$). Farm 5 showed the highest percentage of cows with ears loosely hung down (EP4 = 55.5%), and no cow held the ears upright (EP1 = 0%); farm 4 had the highest percentage of cows with ears held backwards (EP3 = 56.9%); in farm 1, the highest percentage of cows held the ears forwards (EP2 = 65.9%) and upright (EP1 = 13.2%) (Figure 7b).

4. Discussion

Our results show that taking photographs from a distance directly on the farm can be an easy way to collect reliable information regarding eye white and ear posture. This method is quicker and more feasible than video recording. On the other hand, this method does not allow for collecting changes in ear posture that resulted as relevant in some studies in sheep and lambs (e.g., [10,17,24,25]), but whose interpretation provided contradictory results in dairy cows (e.g., [22,23]). We also excluded the collection of asymmetric ear posture, which was frequently found during social isolation in sheep [17]. The decision to take photographs from one random side was due to our aim to find possible associations between eye white and ear posture taken together on the same animal in different contexts. To the best of our knowledge, this is the first attempt to find this association. Regarding on-farm feasibility of eye white, it also has to be remarked that the use of the four-level classification proposed in the present study, with the aid of a telephoto lens (70–300 mm), made this indicator more feasible than using the computerized approach that was proposed by Sandem et al., 2002 [7]. Therefore, this simplified scale proved to be both valid and practical for on-farm assessment.

Our results confirm that cows simultaneously and consistently use different parts of their head to communicate their emotions. Half-closed eyes are mostly associated to ears hanging down, whereas when the eye white is visible, the ears are often held forwards or upright.

Our study was conducted in commercial farms, hence the situations to which cows were exposed were part of their daily life, with the only exception of the ADF. Pasture was the context in which cows showed the lowest arousal (eyes were half-closed in 67.8% of cows) and probably with positive valence (ears hanging down or backwards in 77.3% of cows). Access to pasture has many beneficial effects on the welfare of dairy cows [31], such as a reduction of mortality rate, mastitis, reproductive problems and lameness, and an increased possibility to move, to rest in comfort, to eat preferable food, and to express species-specific behaviors. For these reasons, cows highly prefer outdoor to indoor housing (for review [32]). It is also supposed that sunlight exposure is rewarding and that cows can derive welfare benefits when they are outdoors [32]. However, we also found 37.9% of cows with ear held forwards. This result suggests that pasture may be both a low arousal situation and a stimulating context where cows actively direct their attention to the surrounding.

It is partly surprising that the level of arousal during pasture is even lower than during the sole resting, although the positive effect of the access to pasture is well recognized (half-closed eyes 67.8% versus 45.3%); in addition, we found 8.2% of cows with ear upright during resting, which confirmed that the level of excitement was high in this context for some animals. We can also hypothesize a negative emotional valence expressed by cows that show this ear posture. Resting can be affected by several factors, such as the cubicle: cow ratio, cubicle design (e.g., length, width), housing system (tethered vs loose housing), etc. In farms where the competition for the resting area was high (as in farm 1, cubicle: cow ratio = 0.89), most of the cows had the eye white clearly visible and the ears held forwards and upright (e.g., the highest percentage of ear in upright position was recorded in farm 1: EP1 = 13.2%). We can suppose that the cows were alerted (high arousal) and nervous (negative valence) during social interactions in highly competitive contexts. In farm 3, where cows were in tie stalls, 44.7% of cows showed eye white from barely to clearly visible, 46% of ear held forwards, and 6% of ears in upright position. This husbandry system is likely to be less comfortable than loose housing, as confirmed by several authors [33–36], which was mainly due to the increased risk for lesions and lameness, and to the restriction of voluntary movements and social behavior. We might hypothesize that being tethered in a tie stall is not relaxing for cows: in fact, [36] observed that lying down position was frequently incorrect in tie-stalls, where almost 35% of cows laid partly or completely outside the stall, or showed evident signs of compression or discomfort of the hind part of the body. Hence, we can suppose that the underlying emotional valence expressed by cows was negative, which was possibly due to discomfort rather than to competition. In farm 5, where the cubicle: cow ratio was positive (1.12), and the cows also had access to additional outdoor space, we found the more relaxed cows (low arousal, positive valence), as shown by the highest percentage of half-closed eyes (96.3%) and hanging ears (55.5%). We hypothesize that, when cows live in environments where the competition for resources is limited, the arousal is low and is probably associated to a positive valence. An additional hypothesis is that pasture may have a long-lasting effect on cows' emotional state, which goes beyond the bare time spent outdoors, and leads to a general more positive state even when animals are indoors.

In our research, during feeding cows were particularly excited (34.9% of cows with eyes from barely to clearly visible; 5.8% with ears in upright position), which is in agreement with other studies showing that feeding produces greater arousal than other activities (e.g., rumination [10]). As already discussed for resting, feeding can be a competitive context and emotions can be influenced by several factors, such as feed quality and space at the rack. Cows seem to be more at ease (positive valence) in farms where the feeding space: cow ratio is at least 1, as in farm 4, or 1.72 as in farm 5, that is where the competition is low. This is supported by the highest percentage of eyes from half-closed to normally open (81.8% in farm 4; 81.2% in farm 5) and, in farm 5, also by the high percentage of cows with ears that were held backwards or hung down loosely, which indicate a positive state that was also probably due to the type of feed (ventilated hay and fresh grass, instead of total mixed ration). This result is in line with a study conducted in sheep fed fresh hay that revealed a high proportion of animals with ears in the backward position [17]. Even for feeding, as for resting, we cannot exclude that this positive emotional state is also partly due to the long-lasting positive effect of the access to pasture. In farm 3 (tie stalls), the feeding place: cow ratio was obviously 1. However, 75% of the cows showed eye white from barely to clearly visible, and the highest percentage of cows with ear upright (12.5%) was recorded. We can suppose that the limited possibility to move during feeding and the frustration due to the reduced possibility to solve social conflicts with neighboring cows may represent a stressful situation (negative valence) even when cows have sufficient space to access the feed rack. Differently, in farm 4, where the feeding place: cow ratio is also 1, the ears are frequently held forwards

(54.5%). Studies on goats [37] revealed that ears forwards can be a sign of vigilance: hence, the arousal is high, but valence may be either positive or negative, and expresses attention to the surrounding rather than stress.

ADF elicited the strongest reaction in cows: eye white clearly visible was mostly recorded during the execution of this test (44.8%). As a high percentage of eye white may be a valid indicator of fear in dairy cows [13], we may suppose that, during the execution of the test, cows are experiencing emotions with a negative valence (i.e., fear) and probably a high arousal. Unsurprisingly, during ADF test, most of the cows directed the ear pinna forwards towards the assessor (95.5%), which confirmed that an approaching person could cause a vigilance reaction in cows. Further studies would be interesting in order to match eye white and ear posture scores to the results of the ADF test for each cow, to gather more information regarding the perception of danger and the quality of the human-animal relationship.

Some of the studies reported that eye white can better express the arousal of emotions [23], whereas the valence can also affect ear posture [22]. Even if both indicators can help interpreting emotions in dairy cows in the same direction, the two measures seem complementary, as eye aperture and ear movements can have different meanings for a cow.

5. Conclusions

This research confirms that eye white and ear posture can be promising indicators to be included in the assessment of emotions of dairy cows on farm. However, even though cows show a rather consistent use of eyes and ears to communicate their emotions, the correlation between these two indicators is not completely linear, and a combined use of eye white and ear posture would probably give a more complete and reliable picture of the valence and arousal that were experienced by the animals.

The collection of these indicators in contexts that are routinely experienced by cows during their daily life can contribute to adding information regarding the perception of the environment and the valence and the arousal of emotions in different situations, but the reliability of these indicators should be further investigated, and their feasibility may be improved by collecting data by direct observations.

Author Contributions: Conceptualization, methodology, data analysis: M.B. and S.M.; data collection: A.A.; writing—original draft preparation: M.B.; writing—review and editing, supervision: S.M.

Funding: This research was funded by Santangiolina Latte Fattorie Lombarde Soc. Agr. Cooperativa.

Acknowledgments: We are grateful to Martin Sanna and Luca Bergamini for their technical support for data collection, and to all the farmers who gave us access to their farms.

Conflicts of Interest: The authors declare no conflict of interest.

References

1. Panel on Animal Health and Welfare (AHAW). Statement on the use of animal-based measures to assess the welfare of animals. *EFSA J.* **2012**, *10*, 1–29. [CrossRef]
2. Wemelsfelder, F.; Millard, F.; De Rosa, G.; Napolitano, F. Qualitative behaviour assessment. In *Welfare Quality® Report No. 11—Assessment of Animal Welfare Measures for Dairy Cattle, Beef Bulls and Veal Calves*; Forkman, B., Keeling, L., Eds.; Cardiff University: Cardiff, UK, 2009; pp. 215–224.
3. Sant'Anna, A.C.; Paranhos da Costa, M.J.R. Validity and feasibility of qualitative behavior assessment for the evaluation of Nellore cattle temperament. *Livest. Sci.* **2013**, *157*, 254–262. [CrossRef]
4. Grosso, L.; Battini, M.; Wemelsfelder, F.; Barbieri, S.; Minero, M.; Dalla Costa, E.; Mattiello, S. On-farm Qualitative Behaviour Assessment of dairy goats in different housing conditions. *Appl. Anim. Behav. Sci.* **2016**, *180*, 51–57. [CrossRef]
5. Boissy, A.; Manteuffel, G.; Jensen, M.B.; Moe, R.O.; Spruijt, B.; Keeling, L.J.; Winckler, C.; Forkman, B.; Dimitrov, I.; Langbein, J.; et al. Assessment of positive emotions in animals to improve their welfare. *Physiol. Behav.* **2007**, *92*, 375–397. [CrossRef] [PubMed]
6. Battini, M.; Barbieri, S.; Vieira, A.; Can, E.; Stilwell, G.; Mattiello, S. The Use of Qualitative Behaviour Assessment for the On-Farm Welfare Assessment of Dairy Goats. *Animals* **2018**, *8*, 123. [CrossRef] [PubMed]

7. Proctor, H.S.; Carder, G. Measuring positive emotions in cows: Do visible eye whites tell us anything? *Physiol. Behav.* **2015**, *147*, 1–6. [CrossRef] [PubMed]
8. Descovich, K.A.; Wathan, J.; Leach, M.C.; Buchanan-smith, H.M.; Flecknell, P.; Farningham, D.; Vick, S. Facial Expression: An Under-Utilized Tool for the Assessment of Welfare in Mammals. *ALTEX* **2017**, *34*, 409–429. [CrossRef]
9. Mendl, M.; Burman, O.H.P.; Paul, E.S. An integrative and functional framework for the study of animal emotion and mood. *Proc. R. Soc. B Biol. Sci.* **2010**, *277*, 2895–2904. [CrossRef]
10. Reefmann, N.; Wechsler, B.; Gygax, L. Behavioural and physiological assessment of positive and negative emotion in sheep. *Anim. Behav.* **2009**, *78*, 651–659. [CrossRef]
11. Lambert (Proctor), H.S.; Carder, G. Looking into the eyes of a cow: Can eye whites be used as a measure of emotional state? *Appl. Anim. Behav. Sci.* **2017**, *186*, 1–6. [CrossRef]
12. Sandem, A.I.; Janczak, A.M.; Salte, R.; Braastad, B.O. The use of diazepam as a pharmacological validation of eye white as an indicator of emotional state in dairy cows. *Appl. Anim. Behav. Sci.* **2006**, *96*, 177–183. [CrossRef]
13. Sandem, A.I.; Janczak, A.M.; Braastad, B.O. Short communication A short note on effects of exposure to a novel stimulus (umbrella) on behaviour and percentage of eye-white in cows. *Appl. Anim. Behav. Sci.* **2004**, *89*, 309–314. [CrossRef]
14. Sandem, A.I.; Braastad, B.O.; Bøe, K.E. Eye white may indicate emotional state on a frustration-contentedness axis in dairy cows. *Appl. Anim. Behav. Sci.* **2002**, *79*, 1–10. [CrossRef]
15. Sandem, A.-I.; Braastad, B.O. Effects of cow—Calf separation on visible eye white and behaviour in dairy cows—A brief report. *Appl. Anim. Behav. Sci.* **2005**, *95*, 233–239. [CrossRef]
16. Sandem, A.I.; Braastad, B.O.; Bakken, M. Behaviour and percentage eye-white in cows waiting to be fed concentrate-A brief report. *Appl. Anim. Behav. Sci.* **2006**, *97*, 145–151. [CrossRef]
17. Reefmann, N.; Bütikofer Kaszàs, F.; Wechsler, B.; Gygax, L. Ear and tail postures as indicators of emotional valence in sheep. *Appl. Anim. Behav. Sci.* **2009**, *118*, 199–207. [CrossRef]
18. Tamioso, P.R.; Rucinque, D.S.; Taconeli, C.A.; da Silva, G.P.; Molento, C.F.M. Behavior and body surface temperature as welfare indicators in selected sheep regularly brushed by a familiar observer. *J. Vet. Behav. Clin. Appl. Res.* **2017**, *19*, 27–34. [CrossRef]
19. Tamioso, P.R.; Maiolino Molento, C.F.; Boivin, X.; Chandèze, H.; Andanson, S.; Delval, É.; Hazard, D.; da Silva, G.P.; Taconeli, C.A.; Boissy, A. Inducing positive emotions: Behavioural and cardiac responses to human and brushing in ewes selected for high vs low social reactivity. *Appl. Anim. Behav. Sci.* **2018**, *208*, 56–65. [CrossRef]
20. Reefmann, N.; Bütikofer, F.; Wechsler, B.; Gygax, L. Physiological expression of emotional reactions in sheep. *Physiol. Behav.* **2009**, *98*, 235–241. [CrossRef]
21. Gómez, Y.; Bieler, R.; Hankele, A.K.; Zähner, M.; Savary, P.; Hillmann, E. Evaluation of visible eye white and maximum eye temperature as non-invasive indicators of stress in dairy cows. *Appl. Anim. Behav. Sci.* **2018**, *198*, 1–8. [CrossRef]
22. Lambert, H.; Carder, G. Positive and negative emotions in dairy cows: Can ear postures be used as a measure? *Behav. Process.* **2019**, *158*, 172–180. [CrossRef] [PubMed]
23. Proctor, H.S.; Carder, G. Can ear postures reliably measure the positive emotional state of cows? *Appl. Anim. Behav. Sci.* **2014**, *161*, 20–27. [CrossRef]
24. Guesgen, M.; Beausoleil, N.; Minot, E.; Stewart, M.; Stafford, K.; Morel, P. Lambs show changes in ear posture when experiencing pain. *Anim. Welf.* **2016**, *25*, 171–177. [CrossRef]
25. Vögeli, S.; Wechsler, B.; Gygax, L. Welfare by the ear: Comparing relative durations and frequencies of ear postures by using an automated tracking system in sheep. *Anim. Welf.* **2014**, *23*, 267–274. [CrossRef]
26. Boissy, A.; Aubert, A.; Greiveldinger, L.; Delval, E.; Veissier, I. Cognitive sciences to relate ear postures to emotions in sheep. *Anim. Welf.* **2011**, *20*, 47–56.
27. Bellegarde, L.G.A.; Haskell, M.J.; Duvaux-ponter, C.; Weiss, A.; Boissy, A.; Erhard, H.W. Face-based perception of emotions in dairy goats. *Appl. Anim. Behav. Sci.* **2017**, *193*, 51–59. [CrossRef]
28. Schmied, C.; Waiblinger, S.; Scharl, T.; Leisch, F.; Boivin, X. Stroking of different body regions by a human: Effects on behaviour and heart rate of dairy cows. *Appl. Anim. Behav. Sci.* **2008**, *109*, 25–38. [CrossRef]

29. Coulon, M.; Nowak, R.; Peyrat, J.; Chandèze, H.; Boissy, A.; Boivin, X. Do Lambs Perceive Regular Human Stroking as Pleasant? Behavior and Heart Rate Variability Analyses. *PLoS ONE* **2015**, *10*, e0118617. [CrossRef] [PubMed]
30. Windschnurer, I.; Schmied, C.; Boivin, X.; Waiblinger, S. Assessment of Animal Welfare Measures for Dairy Cows, Beef Bulls and Veal Calves. In *Welfare Quality® Reports, Vol. 11*; Forkman, B., Keeling, L., Eds.; School of City and Regional Planning Cardiff University: Cardiff, UK, 2009; pp. 137–152.
31. European Food Safety Authority. Scientific report on the effects of farming systems on dairy cow welfare and disease. *EFSA J.* **2009**, *7*, 1143r. [CrossRef]
32. Arnott, G.; Ferris, C.P.; O'connell, N.E. Review: Welfare of dairy cows in continuously housed and pasture-based production systems. *Animal* **2017**, *11*, 261–273. [CrossRef]
33. Plesch, G.; Broerkens, N.; Laister, S.; Winckler, C.; Knierim, U. Reliability and feasibility of selected measures concerning resting behaviour for the on-farm welfare assessment in dairy cows. *Appl. Anim. Behav. Sci.* **2010**, *126*, 19–26. [CrossRef]
34. Rushen, J.; Haley, D.; de Passillé, A.M. Effect of Softer Flooring in Tie Stalls on Resting Behavior and Leg Injuries of Lactating Cows. *J. Dairy Sci.* **2007**, *90*, 3647–3651. [CrossRef] [PubMed]
35. Mattiello, S.; Klotz, C.; Baroli, D.; Minero, M.; Ferrante, V.; Canali, E. Welfare problems in alpine dairy cattle farms in Alto Adige (Eastern Italian Alps). *Ital. J. Anim. Sci.* **2009**, *8*, 628–630. [CrossRef]
36. Mattiello, S.; Arduino, D.; Tosi, M.V.; Carenzi, C. Survey on housing, management and welfare of dairy cattle in tie-stalls in western Italian Alps. *Acta Agric. Scand. Sect. A Anim. Sci.* **2005**, *55*, 31–39. [CrossRef]
37. Briefer, E.F.; Tettamanti, F.; McElligott, A.G. Emotions in goats: Mapping physiological, behavioural and vocal profiles. *Anim. Behav.* **2015**, *99*, 131–143. [CrossRef]

© 2019 by the authors. Licensee MDPI, Basel, Switzerland. This article is an open access article distributed under the terms and conditions of the Creative Commons Attribution (CC BY) license (http://creativecommons.org/licenses/by/4.0/).

Article

Investigation of a Standardized Qualitative Behaviour Assessment and Exploration of Potential Influencing Factors on the Emotional State of Dairy Calves

Marta Brscic [1], Nina Dam Otten [2], Barbara Contiero [1] and Marlene Katharina Kirchner [2,*]

1. Department of Animal Medicine, Production and Health, University of Padova, 35020 Legnaro (PD), Italy; marta.brscic@unipd.it (M.B.); barbara.contiero@unipd.it (B.C.)
2. Institute of Animal Welfare and Disease Control, Dep. Veterinary and Animal Sciences, University of Copenhagen, 1870 Frederiksberg C, Denmark; nio@sund.ku.dk
* Correspondence: marlene.kirchner@vier-pfoten.org; Tel.: +43-15455020195

Received: 2 September 2019; Accepted: 30 September 2019; Published: 2 October 2019

Simple Summary: Although welfare states of dairy calves are of public and scientific concern, no standardized protocol exists to assess the emotionality of these animals. Therefore, this study aimed at investigating and establishing a calf-specific term list for Qualitative Behavior Assessment (QBA), a technique that is already validated for assessing emotional states in many animal species. The statistically supported results showed that agreement can be reached among observers, terms showed varied results across farms, and evaluated emotional states could be linked to some explaining farm factors. Overall, results showed that calves have a neutral emotional state and profit from certain farm factors. However, we conclude that the assessment should be more widely used to gain more insight into calves' welfare states and how their emotional state can be improved to a positive one.

Abstract: Assessing emotional states of dairy calves is an essential part of welfare assessment, but standardized protocols are absent. The present study aims at assessing the emotional states of dairy calves and establishing a reliable standard procedure with Qualitative Behavioral Assessment (QBA) and 20 defined terms. Video material was used to compare multiple observer results. Further, live observations were performed on 49 dairy herds in Denmark and Italy. Principal Component Analysis (PCA) identified observer agreement and QBA dimensions (PC). For achieving overall welfare judgment, PC1-scores were turned into the Welfare Quality (WQ) criterion 'Positive Emotional State'. Finally, farm factors' influence on the WQ criterion was evaluated by mixed linear models. PCA summarized QBA descriptors as PC1 'Valence' and PC2 'Arousal' (explained variation 40.3% and 13.3%). The highest positive descriptor loadings on PC1 was Happy (0.92) and Nervous (0.72) on PC2. The WQ-criterion score (WQ-C12) was on average 51.1 ± 9.0 points (0: worst to 100: excellent state) and 'Number of calves', 'Farming style', and 'Breed' explained 18% of the variability of it. We conclude that the 20 terms achieved a high portion of explained variation providing a differentiated view on the emotional state of calves. The defined term list proved to need good training for observer agreement.

Keywords: calves; emotional state; organic; farm size; term list

1. Introduction

Positive indicators in animal welfare assessment schemes are still limited although, nowadays, essential in many ways. Qualitative Behavioral Assessment (QBA) focuses on animals' demeanor and body language to identify the underlying emotional state. QBA is one possibility which is frequently used to assess animals' emotional states on-farm and in experiments, using either a free choice profiling (FCP) [1] or term lists as behavioral descriptors [2]. Validity aspects of QBA were investigated in various animal species and situations such as the ability to picture anxiety in pigs [3], in relation to farm factors in veal calves [4], or stress levels during transport in sheep [5]. Additionally, reliability was proven in some studies and species, such as intra-day variation of QBA in dairy cows [6]. After investigations on observer reliability in dairy cows, fattening cattle, and veal calves [2], QBA was implemented as a measure for the criteria 'positive emotional state' (WQ-C12) in the Welfare Quality (WQ) assessment protocols for cattle, pigs, and chicken. QBA was used to address the criteria 'positive emotional state' (WQ-C12) [2] with specified terms for the respective livestock species, category, or types. In parallel to the development of these standardized lists in WQ, mathematical procedures were set in place to summarize a Principal Component Score (PC1). The potential range of this PC1 underwent an expert-based evaluation leading subsequently to a welfare interpretation of the QBA result into the WQ-C12, achieving points on a welfare state scale from 0 (worst) to 100 (excellent). Whenever establishing a QBA for new species, studies on reliability between observers and association with other factors, such as housing conditions or health state, were part of the preliminary studies. These studies aimed at validating the assessment method by looking at variation in the sample and at the potential of QBA to discriminate between each husbandry system. Several studies assessing and investigating reliability and associated factors of QBA with term lists, and additionally evaluating welfare in terms of the emotional state, were published for donkeys [7], cattle [8,9], pigs [10], and sheep [11], in particular. The advantages of QBA procedures with standardized term lists and calculations are the easier application and a straightforward interpretation of results by experienced assessors and experts, in particular when expressed within the welfare evaluation of the WQ-C12. The latter is given by the integrative handling of the collected QBA data, summing 20 terms in WQ-C12, allowing an easy calculation of a representative score at herd level and enabling an easier comparison between farms, systems, observers, and, furthermore, the identification of influencing factors for good and poor results. Consecutively, this might help to relate different management and farming practices to QBA results and add distinct information to a holistic welfare assessment [12] in dairy calves made by clinical examination and quantitative behavior observations as shown earlier for beef cattle [13]. Recently, such a QBA criterion score (WQ-C12) provided information transferrable to consumers, as well as valuable feedback to the farmer wishing to implement improvements [8,14].

In the past years, the welfare assessment in dairy cow systems focused only on lactating cows, neglecting the fact that also young calves and heifers are kept in large numbers on-farm and excluding them from welfare assessments, with a minor exception by Gratzer et al. [15]. Not only is the female offspring of particular high value for the farmer and the production systems' economy, but at the same time the number of affected animals itself implies that they should be included in a welfare assessment, being representative of the farm and production system as a whole. Furthermore, in common dairy farming practice calves and cows are still forcedly separated after a few hours or days, both in organic and conventional systems. It could therefore be interesting to investigate the welfare state of the calves from a holistic point and emotional state, particularly seeing today's need for alternatives to the early cow–calf separation [16]. In Norway, [17] the human–animal relationship in dairy calves was investigated using 31 QBA terms due to a missing standard procedure for dairy cows, fattening cattle, and veal calves [12] to describe dairy calves' body language in handling situations by stockpeople.

Calf husbandry practices in dairy farming follow some basic characteristics across Europe, such as the early (0–48 h) separation from the cow, the initial single or double penning followed by group housing at a later age, and very similar feeding strategies. However, at farm level the two included countries, Italy and Denmark, show great variation in structure and management regarding the outdoor or indoor housing of the calves, the access to pasture, lying comfort, health surveillance, the size of the farm both in number of animals and revenue, the kept breeds, or the available manpower. Some of these farm factors could potentially influence the emotional state of the calves and could be used in an analysis together with QBA results. Herd size is one of the interesting factors as this shows the greatest variation both within and between the two countries. The intensification of the dairy production over the past decades has left the Danish herds with an average herd size of 167 cows in

2015 compared to an average of 65.9 cows in 2000, rendering the larger herds with increased challenges regarding mortality [18] similar to their Nordic neighbors [19,20]. Additionally, herd size has also been associated with other management practices such as the choice of grazing or organic production status in smaller farms [21], an issue also evident in the Italian dairy production [22]. Larger production sites require a larger number of employees. In some farms, however, this only implies a greater number of animals per stockperson, which could negatively affect the welfare of the entire herd as discussed by Simon et al. [23]. Within the last decades, a vast number of non-natives being employed on Danish farms for varying time periods has led to a high turnover in manpower, challenging management in terms of language and cultural barriers and possibly different attitudes towards calf handling and rearing. To our knowledge, no previous studies have investigated the potential influence of these farm factors on the emotional state of calves. Likewise, studies evaluating observer agreement on QBA video material for the evaluation of a prespecified term list in dairy calves has not been published so far. Furthermore, information on emotional states on calves in dairy herds, especially in terms of aggregated scores (WQ-C12) at farm level, was desirable for future purposes.

Therefore, this study aimed at investigating a QBA procedure, including 20 terms, by extrapolating the main dimensions of the calves' emotional states achieved by Principal Component Analysis (PCA) and testing reliability by means of comparing observer agreement on video material. Further, we performed the QBA aggregation procedure and integrated score calculation to achieve Welfare Quality criterion scores (WQ-C12) at farm level. Finally, we were investigating certain farm factors potentially related to the emotional state of dairy calves for identifying explanatory variables on WQ-C12.

2. Materials and Methods

2.1. QBA Procedure

The WQ protocol for calves and heifers developed earlier [16] and the proposed QBA description was used for the on-farm assessments of emotional state. The proposed QBA was a term list with 20 descriptors, 'Active, Relaxed, Uncomfortable, Calm, Content, Tense, Enjoying, Indifferent, Frustrated, Friendly, Bored, Positively occupied, Inquisitive, Irritable, Nervous, Boisterous, Uneasy, Sociable, Happy, Distressed', which was used to score behavior after an observation period of 20 min. Each term was scored on a 125 mm continuous scale (Visual Analogue Scale), where left represented the 'minimum' and right the 'maximum' point, by crossing each scale at a certain point fitting the observation. A very left position on the scale or a zero indicated that the expressed quality of the specific behavior/term was "entirely absent in any of the animals seen", whereas the 'Maximum' at 125 mm stood for "the expressed quality of the specific behavior/term was constantly obvious across all animals seen during the observation" as was relevant in the case of herd observations [2]. Additionally, further considerations had to be integrated in the scoring by the observer, as the aim was also to imagine the respective endpoints of the scale as what was possible as an expression of a behavioral quality for the respective species for the given age and sex. Hence, the term "Inquisitive" would be at a different expected level in adult cows compared to calves. Training and testing of observers was performed to ensure calibration and the correct use of the scale for QBA. Initially, this included exchange of experiences amongst observers on different impressions from different husbandry systems or circumstances leading to minimum or maximum points on certain behavioral expressions, illustrating for their colleagues' potential magnitudes and enhancing understanding of the potential dimensions.

2.2. Video Sessions and Observers

Reliability testing of the chosen QBA term list was done by including eleven trained observers (7 females, 4 male) to a test panel which was asked to score QBA in 20 videos showing calves, in one go, with a length of 1–2 min each. All observers had a higher education at a university level in the field of Animal Science or Veterinary Medicine from across Europe, with both males and females being represented. All participants had a minimum experience of a week-long course on QBA in cattle (cows and calves), including on-farm and video practice, and score calculations with the WQ-QBA lists for dairy cattle and dairy calves. Prior to the video session, all the terms were discussed and calibrated again amongst all observers according to the definitions that can be found in Table 1. All observers scored the videos on QBA paper sheets with the above-mentioned term lists and transferred their scores (in mm) to a provided electronic data sheet.

Table 1. QBA fixed-term list (20 terms) with definition of terms, loadings of Principal Component 1 and 2 (PC1 and PC2) and extracted weights for the simplified score aggregation.

Terms (Factors)	Definition of Term	PC1	PC2	Weights (Estimate)
(Intercept) = 'constant'				−2.031
Active	Engage in an activity in a conscious manner, regardless of motor activity; can be resting, while attentive	0.39	0.63	0.001792
Relaxed	Body language is at ease, animals seem not stressed	0.74	0.03	0.003648
Uncomfortable	Animals display physical or mental discomfort in the given situation, may show attempts to avoid source of discomfort	−0.67	0.30	−0.006389
Calm	Even-tempered, still and quiet in the performance of activities	0.62	−0.19	0.003380
Content	Animals express overall contentment with their life situation and seem mentally balanced, in control of their constitution	0.91	−0.08	0.004339
Tense	Rigid postural or /and facial expression, muscle strains visible, stiff body posture or movements	−0.56	0.68	−0.009337
Enjoying	Express satisfaction in the given situation and occupation	0.76	0.21	0.003411
Indifferent	Animals are aware of their environment and stimuli, but do not engage in activities or react	−0.61	−0.36	−0.003649
Frustrated	Motivation cannot be satisfied, may lead to compulsive or replacement behavior	−0.74	0.48	−0.009898
Friendly	Animals display friendliness towards other animals and humans	0.65	0.31	0.003293
Bored	animals are idling or active without any purpose, no motivation detectable to engage in any kind of activity (Results of prolonged lack of stimuli)	−0.55	0.18	−0.003098
Positively occupied	Animals actions express that they "like to do what they do" in a given situation	0.80	0.09	0.003530
Inquisitive	Desire to perform active investigation of surroundings or conspecifics on the hunt for new experiences and stimuli	0.54	0.35	0.002364
Irritable	animals are easily upset and irritated or agitated	−0.38	0.45	−0.006465
Nervous	Elevated level of arousal and vigilance, might be combined with restlessness and body movements	−0.18	**0.72**	−0.002303
Boisterous	Heedless, reckless behavior without any sign of aggression	0.24	0.17	0.001505
Uneasy	Physically or mentally troubled, long-term state of discomfort	−0.72	0.00	−0.006855
Sociable	Actively seeking for social engagement	0.50	0.30	0.002185
Happy	Displaying excitement, joy and pleasure in a given situation	**0.92**	−0.02	0.004075
Distressed	Comprised adaptability resulting in incapability of action; animals resign, withdraw completely; close to death	−0.60	−0.41	−0.009291

Highest loading for each term is typed in bold.

2.3. QBA Application On-Farm, Observers, and Farms

Before the on-farm assessments started, four observers familiarized themselves with the terms of the above-mentioned list and their descriptions (Table 1), ensuring they understood the terms and used them in a similar context. Furthermore, the four observers, all female and veterinarians, applying the QBA were all previously trained for WQ in dairy cattle and welfare assessment of dairy calves including the new QBA system, respectively. They achieved sufficient agreement with a silver standard (previously trained and certified person) and each other (r > 0.7). For logistical reasons and limited resources of the study, each farm was visited by one observer only, observers assessed in their respective affiliated country. For the same reasons, farms were recruited by the observers themselves in their own countries, firstly dependent on farmers' willingness to join voluntarily, and secondly observers tried to involve a high variation of farm sizes, breed, farming styles, and locations in order to potentially achieve a large variation of emotional states in calves. Therefore, the sample of farms ended up being a conveniently chosen one, an expected situation in on-farm studies, as nothing else was feasible. Nonetheless, it covered a wide range of different husbandry conditions for the calves, and therefore expectedly also a variety of potentially possible QBA results.

The on-farm assessments were carried out in dairy cow farms in Denmark (40) and in Italy (9), 31 conventional and 18 organic farms size varying from 106–608 lactating cows with an average of 48.2 calves per farm. Calves were mainly housed single or pairwise in the first month and later on moved to group housing. Breeds present on farms were either Holstein (25), other milk types such as Jersey or Brown Swiss (13), or mixed, meaning mixed breeds and/or mixed herds (11). In 13 of the farms, weaned calves grazed on pasture, farms keeping males had on average 12 bulls, and the number of full-time equivalent workers ranged from 1 to 3.5. The assessment was carried out in the early morning (around milking time for the cows, before morning feeding for the calves), as the first evaluation upon arrival at the farm. The calves at herd level were observed for 20 min in total, including all animals (male and female) aged 0–180 days present at the time of inspection. If animals were housed in several groups, the observation time between groups was split with a maximum of 8 observation points. In principle, animals were observed at group level even if housed individually or pairwise, in as large observational segments as possible and visible. After finishing the observations, the observers turned away from the animals, i.e., walked away from the animals to not observe any further, and made their scoring on a paper version of QBA.

2.4. Statistics

The data collected on-farm were analyzed using a Principal Component Analysis (PCA). The analysis was based on a correlation matrix, without rotation, and two components were extracted (after pre-exploration of an unlimited number of components and selection of eigenvalues >1 adjusted by the explained variances to avoid under/over factoring). Analyses were performed in R [24] using the libraries rcmdr, psych, Deducer, DeducerExtra. The dataset was sent to an external researcher, experienced with QBA and PCA, revealing the same PCA results and Graph of terms with a different statistical program (Minitab) to reassure correct methodology.

Accordingly, scores attributed to the calves by the 11 observers during the video's administration were submitted to PCA analysis using a correlation matrix with no rotation. The PCA scores attributed to the 20 videos on the first two main Principal Components were tested for interobserver reliability using Kendall Correlation Coefficient W. Kendall W values can vary from 0 (no agreement at all) to 1 (complete agreement), with values higher than 0.6 showing substantial agreement. Subsequently, the interobserver reliability for each descriptor separately was calculated using the intraclass correlation coefficient (ICC).

2.5. Procedures with the On-Farm QBA Scores

The PC1-score, obtained from the PCA for each farm, was analyzed as a dependent variable in a linear regression model using the 20 terms as predictors to estimate the weights (reported as "estimates" in Table 1) according to the approach by Budaev [25]. The WQ criterion score (WQ-C12), defined in the WQ cattle protocol [12], was calculated using the following formula:

$$WQ - C_{12} = constant + \sum_{k=1}^{20} w_k N_k \quad (1)$$

where N_k is the value (in mm) obtained by a farm for a given term k, w_k and the weight (= estimate, given in Table 1) attributed to a given term k 'constant', a fixed value for each farm (= intercept, given in Table 1)

The WQ-C12 was used for the interpretation of the calves' welfare state. This index of the 'Emotional state' was analyzed applying a linear model and considering given farm characteristics and managerial choices as predictors (factors). The factors that were constant for every farm with no variation (e.g., age at separation from cow, if born in a calving box) or with too much missing information (>80%, e.g., bedding type of calves) were dropped out. The following factors could be considered for the analysis: housing outdoors or indoors of young calves (HOUSING_Y) and older calves (HOUSING_O) respectively, if weaned calves were on pasture or not (WEANED_PASTURE), the number of calves (CALVES_NO), cows (COWS_NO), young heifers (NO_YOUNGSTOCK) and bulls (NO-BULLS) on the farm, the prevalent cattle breed (BREED), the farming style as 'organic' or 'conventional' (FARMSTYLE), and how much manpower was available (MANPOWER). After pre-elimination of correlated factors, a back/forward procedure based on AIC revealed the best fitting model. Residuals were graphically checked using the function qqplot in R.

3. Results

3.1. PCA for Video and On-Farm Data

Analysis of the video data looking into the agreement of 11 different observers, revealed good overall agreement (Cronbach's Alpha 0.83) between observers on the QBA for dairy calves and an ICC of 0.83. Overall explained variance of the PCA was 29.15% for PC1 and 16.34% for PC2, with an eigenvalue of 5.83 and 3.39, respectively. Agreements between observers for each descriptor were good for 14 terms (Cronbach's Alpha 0.74–0.97) and moderate for five terms (Cronbach's Alpha 0.46–0.64), the term 'distressed' couldn't be analyzed (further details in Table 2).

Table 2. Cronbach's Alpha, intraclass correlation coefficient (ICC), and number of agreements between observers for each descriptor used by the observers to assess the 20 videos.

Descriptor	Cronbach's Alpha	ICC	No. Observers	Significance
active	0.89	**0.84**	11	<0.001
relaxed	0.74	0.59	11	<0.001
uncomfortable	0.77	**0.74**	11	<0.001
calm	0.84	**0.75**	11	<0.001
content	0.82	**0.75**	11	<0.001
tense	0.83	**0.78**	11	<0.001
enjoying	0.90	**0.88**	11	<0.001
indifferent	0.46	0.29	11	0.021
frustrated	0.49	0.44	11	0.011
friendly	0.88	**0.86**	11	<0.001
bored	0.87	**0.80**	11	<0.001
positively occupied	0.93	**0.90**	11	<0.001
inquisitive	0.64	0.54	10	<0.001
irritable	0.58	0.54	10	0.002
nervous	0.80	**0.76**	11	<0.001
boisterous	0.97	**0.97**	11	<0.001
uneasy	0.52	0.50	11	0.007
sociable	0.91	**0.90**	11	<0.001
happy	0.86	**0.75**	11	<0.001
distressed	-	-	11	-[1]

[1] couldn't be analyzed as eight observers scored 0mm for the term. ICC values ≥0.60 are bold typed.

PCA for the on-farm data summarized QBA descriptors on two main components, PC1 and PC2, with eigenvalues of 8.065 and 2.662, explaining 40.3% and 13.3% of the variation, respectively. Farm PC1 scores ranged between −2.67 and 1.49 and PC2 scores from −2.17 to a maximum of 2.42. The PC1 and PC2 loadings for the terms can be seen in detail in Table 1, and their distribution in the four quadrants in Figure 1.

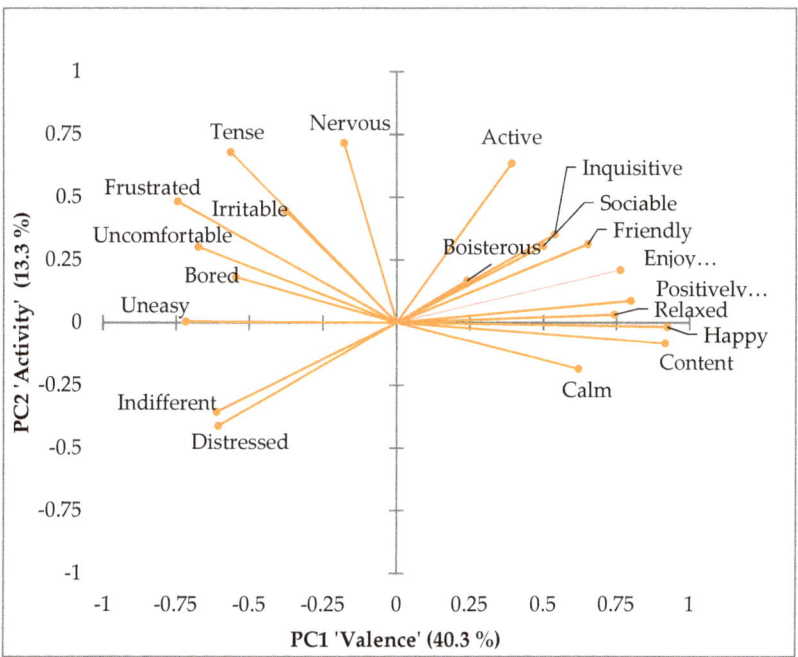

Figure 1. Two-dimensional loading plot of the Principal Component Analysis (PCA) results showing the distribution of all 20 terms describing the quality of calf behavior on PC1 (Valence) and PC2 (Arousal) as vectors.

Descriptors' loadings with the highest positive values on PC1, summarized as 'Valence', were: 'Relaxed', 'Calm', 'Content', 'Enjoying', 'Friendly', 'Positively occupied', 'Inquisitive', 'Sociable', and 'Happy', loaded with a value greater or equal to 0.5. Furthermore, descriptors 'Uncomfortable', 'Tense', 'Indifferent', 'Frustrated', 'Bored', 'Uneasy', and 'Distressed', with a value smaller or equal to −0.56, represented the opposite direction. PC2, described as 'Activity', revealed 'Active', 'Tense', 'Frustrated', and 'Nervous' with values greater or equal to 0.48 as the most positive terms, and 'Calm', 'Indifferent', and 'Distressed' with values smaller or equal to −0.36 on the negative dimension (Table 1). On an individual level, farm results showed great dispersion across the two dimensions, as can be seen in the PCA graph 'farm results' in Figure 2, thus indicating that a discrimination of farms with two dimensions is satisfactory and sensitive enough to picture different emotional states of calf herds.

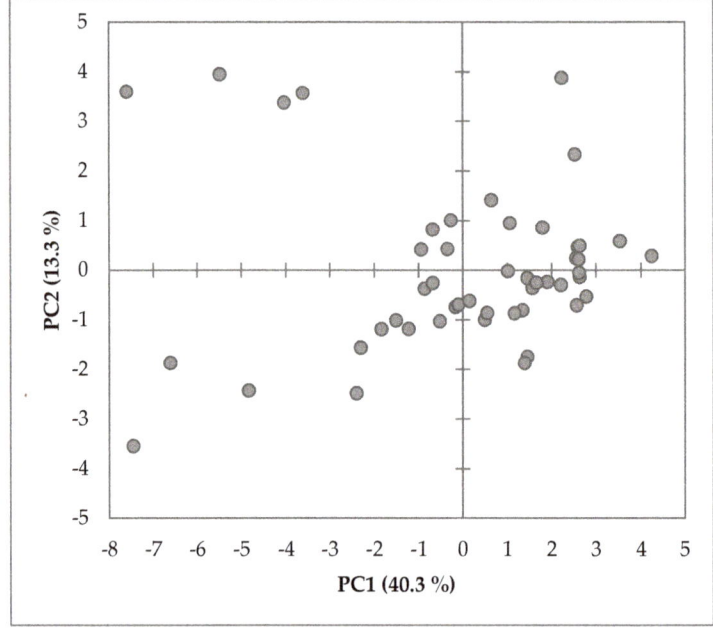

Figure 2. Two-dimensional loading plot of the PCA showing the distribution of the individual farm results of all 49 farms for PC1 and PC2.

3.2. Welfare Quality Scores for the Emotional State of Calves and Associated Factors

The PC1 for each farm, translated to the WQ-criterion score (WQ-C12) gave a mean of 51.1 ± 9.0 points (0 points = worst to 100 points = excellent situation), indicating neutral states (on average) in the involved farms. However, as farm scores ranged from 28.84 to 66.19 points, the average was not representing the actual emotional state for all calf herds included in the study. Results of investigating between-farm variance by modeling WQ-C12 results with certain farm factors revealed that CALVES_NO, FARMSTYLE, and BREED stayed in the model, explaining WQ-C12 best (p-value < 0.05; $r^2_{adjusted}$: 0.18). Farms following the organic style achieved significantly higher estimates (+6.77; p < 0.01). In terms of farm size, every additional calf was revealing a significant, but small, positive effect (+0.05; p < 0.05). For 'BREED', effects were not significant and in our study. Holsteins had the highest (+3.75; p = 0.21) and Jersey the lowest estimates (−1.75; p = 0.61).

4. Discussion

4.1. PCA and Achieved Dimensions

Firstly, this study aimed at investigating the dimensions created by Principal Component Analysis (PCA) using 20 terms describing different quality of behavior in dairy calves. Like other studies applying this procedure, PC1 was found to summarize terms describing 'Valence' and PC2 'Activity', comparable to PC1 'Mood' and PC2 'Activity' for donkeys [7], cows [6], or pigs [10]. The distribution of certain terms across the four quadrants, and their meaningful loadings (>0.5) on PC1 and PC2 [25], also fitted the expressed behavior. Amongst others, the term 'Frustrated', loading very negatively on 'Valence' but positively on 'Activity', was well aligned with the expectations when picturing the behavioral modulation in young calves using this set of terms. Furthermore, the defined components can be seen as a good summary of the overall emotional state, as terms that are close in their expressed quality also go well together in groups across the quadrants, such as 'Happy' and 'Content', or 'Indifferent' and 'Distressed'. There was only one term, 'Boisterous', not loading substantially on one of the two axes, indicating that it is not substantially needed to describe the emotional states of calves. Therefore, it could be argued to exchange it with another term or leave it out and redo the analysis. On the other hand, there might be farms in the future where this term will show a greater response, as it should be seen in the light of our case-to-variable ratio being 2.25, lower than reported from other studies by Budaev [25]. Besides that, breed differences in temperament might be causing this vaguer response. Coming from different countries and with different linguistic backgrounds, the team was communicating and scoring QBAs in English, giving room for slightly different interpretations of the term 'Boisterous', especially in young animals. This might be mitigated in the future by the intensive use of written QBA descriptors and definitions, such as those published for donkeys by Minero et al. [7] also seen in Table 1 in this paper. This approach would probably be helpful but cannot replace training and live agreement between observers. However, as shown by the weights that were assigned to 'Boisterous', it did not have a major impact on WQ-C12 and should be kept in the term list until further studies and more data are available. A similar argument could be done for the negative term 'Distress', which is a severe condition and thus was not recorded for the majority of observers in this study but would be very useful in case QBA is applied in extremely poor welfare conditions or discriminating ones.

4.2. Agreement of Observers in Video Study

For most of the terms, it seems that the current training regime and previous knowledge of QBA amongst the observers was sufficient for a good agreement. However, when looking at the terms with low agreement, which to a large extent described negative states such as 'Indifferent', 'Frustrated', 'Irritable', or 'Uneasy', it seems that observers had recognized different valences. Disagreement can have two reasons—different perceptions or different scorings. With regard to video sessions, well known limitations are the quality of the footage and the restricted viewing angle and overall impression towards the animal an observer has. In short clips, it might be difficult for some observers to focus immediately on the animals' demeanor. However, short clips are preferred to avoid observers suffering from fatigue throughout a long video session, since a minimum number of videos is necessary for the analysis.

A further source of disagreement is a very low or disguised level or absence of a certain term throughout videos, or unclear demarcation of terms. When confronting the observers with the inter-observer reliability of the video session, there were some remarks that some video footage was not good enough to see a lot of the body language of the animals. This might have caused the disagreement and highlights the importance of the footage quality. At the same time, it underlines the importance of training the second source of errors occurring when the scoring is not performed properly—what cannot be seen in the animals should also not be scored (or estimated instead). A general point in scoring QBA is the reference for the line from min to max that some observers build up with experience, to classify the present animals with respect to a certain term. In this video session this was maybe difficult for some observers, in particular if there was limited experience with a huge variety of emotional states in animals, and it might be that this aspect needed more attention in the training material.

For proper scoring, observers also needed to be trained for the occasion that there might be different animals showing different graded valences. To what extent observers were able to implement this cannot be analyzed, and it therefore remains a potential source of the extent of disagreement in the above-mentioned terms. However, as the main amount of terms was scored with good agreement, it has been shown that agreement in principle can be trained and is possible. Future training sessions should therefore focus on terms with so far lower agreement and provide excellent video footage.

4.3. Emotional State of Dairy Calves on the Farms

As stated earlier, this was the first study for dairy calves aggregating the PC1 score to a WQ-C12, and therefore scaling QBA results from 0 (poor) to 100 (excellent) points on a welfare scale. Therefore, we lack direct comparison. In any case we can say the farms in this study reached an average neutral to slightly positive welfare state, based on the expert opinions of Welfare Quality [12]. Comparing PC1 scores of our study with veal calves from 24 Italian farms [4] showed a larger range for the PC1 (−5.08 to 3.88). Translated to WQ-C12, this would correspond to 13.6 to 86.9 points and an average of 52.5, pointing towards a neutral to positive emotional state. Another QBA in 63 beef bull farms with three assessments per farm, using almost the same terms as for the dairy calves, was revealed average PC1 scores ranging from −4.8 to 4.8. This corresponds to 15.2 to 93.2 points for WQ-C12, with an average of 48 points, stating a neutral emotional state in beef bulls [8]. Despite having a larger sample size in the present study than the compared veal studies, but lower sample size than the beef studies, a lower range in WQ-C12 was found. This might be explained by the fact that Danish dairy calf husbandry, which was finally dominant in this study, was more uniform and scorings were therefore more homogenous than in Italian veal calves and beef bulls from Austria, Germany, and Italy [4,8]. However, QBA was still able to discriminate between farms amongst the dairy calf herds of this study, furthermore corresponding to certain farm factors, which is arguing for a large enough variation and sensitivity of the defined terms list.

4.4. Associated Factors on the Welfare Quality Scores for Emotional State of Calves

Finally, as mentioned before, we were investigating certain explanatory farm factors and their potential association with the emotional state of dairy calves at farm level. Positive effects on the emotional state of dairy calves were found by herd size (i.e., number of calves), organic production style, and breed. However, breed had no significant effect in the univariable analyses. Nonetheless, breed can be seen in relation to herd size, as the larger Danish herds most commonly are Holstein herds, which was also reflected by the included study herds. Additionally, a large number of Jersey bull calves are culled shortly after birth, leaving only heifer calves on the dairy farms for assessment. The associations found in the present study are well aligned with other findings, due to the effects of organic production setup. Pairwise housing and group housing of calves are mandatory in organic farms, which lead to decreased fear levels [26,27]. Furthermore, enhancement in social interactions is aiding calves during the feed uptake in the transition period at weaning [28,29].

Studies investigating potential risk factors for an aggregated emotional state in cattle farms are rare to our knowledge. Brscic et al. [4] found differences in the descriptive terms 'Active' and 'Lively' according to the housing form of veal calves at the age of 3 weeks (single vs. group housed). In contrast to that, the present study could not find an influence of the type of housing in our study, neither in very young calves nor older ones. Ellingsen et al. [17] analyzed effects of four different handling styles of stockperson ('Calm/Patient', 'Dominating/Aggressive', 'Positive interactions', and 'Insecure/Nervous') and succeeded in proving an influence on positive and negative moods of dairy calves. Although stockpersons' handling in our study was not observed due to study limitations, it would be valuable to include this in future studies on risk factors of emotional state and would probably lead to a higher explained variance in the model. The aspect of including stockmanship was also supported by the study from Norwegian colleagues [30], who investigated effects of farm factors and stockpersons' handling style on QBA in goats. Findings included that the positive attitude towards petting goats significantly related negatively to the QBA term 'Aggressive' and positively to the term 'Inquisitive/Interested'. In this study, we have also made farmer questionnaires with some of the farmers regarding their attitudes towards Animal Welfare, which is currently under investigation and will be part of our next study.

5. Conclusions

We conclude that with the 20 terms used, it is possible to achieve good overall observer agreement with trained persons. However, analysis pointed out that some terms would profit from improved video training of observers for a common understanding. We observed a relatively high portion of explained variation (54.0%) on PC1 ('Valence') and PC2 ('Activity') providing a differentiated view on the emotional state of calves. WQ-C_{12}-scores were pointing at a neutral to positive emotional state, of which 18% of the variance could be related to number of calves on a farm, farming style, and prevalent breed, leaving enough potential for further investigations into which factors on farm might influence the emotional state.

Author Contributions: Conceptualization, M.K.K.; methodology, M.K.K., M.B., N.D.O., B.C.; investigation, M.K.K., M.B., N.D.O.; data curation, M.K.K., B.C.; writing—original draft preparation, M.K.K.; writing—review and editing, M.K.K., M.B., N.D.O., B.C.; visualization, M.K.K., B.C.

Funding: This research received no external funding.

Acknowledgments: We would like to thank our Italian colleague Isabella Lora for helping with the data collection, Francoise Wemelsfelder for consultancy on the PCA-calculation, the 11 volunteering observers and last not least, the farmers and animals for their excellent collaboration. Earlier versions of the work and PCA-weights extractions were presented on the ISAE-Conference in Edinburgh July 2016 and the ECAWBM- Conference in Cascais October 2016.

Conflicts of Interest: The authors declare no conflict of interest.

References

1. Wemelsfelder, F.; Lawrence, A.B. Qualitative Assessment of Animal Behaviour as an On-Farm Welfare-monitoring Tool. *Acta Agric. Scand. Sect. A Anim. Sci.* **2001**, *51*, 21–25.
2. Wemelsfelder, F.; Millard, F.; De Rosa, G.; Napolitano, F. Qualitative Behaviour Assessment. In *Welfare Quality Reports No. 11*; Forkman, B., Keeling, L.J., Eds.; Welfare Quality Consortium: Lelystad, The Netherlands, 2009; pp. 215–224.
3. Rutherford, K.M.D.; Donald, R.D.; Lawrence, A.B.; Wemelsfelder, F. Qualitative Behavioural Assessment of emotionality in pigs. *Appl. Anim. Behav. Sci.* **2012**, *139*, 218–224. [CrossRef] [PubMed]
4. Brscic, M.; Wemelsfelder, F.; Tessitore, E.; Gottardo, F.; Cozzi, G.; Van Reenen, C. Welfare assessment, correlations and integration between a Qualitative Behavioural Assessment and a clinical/health protocol applied in veal calves farms. *Ital. J. Anim. Sci.* **2009**, *8*, 601–603. [CrossRef]
5. Wickham, S.L.; Collins, T.; Barnes, A.L.; Miller, D.W.; Beatty, D.T.; Stockman, C.; Blache, D.; Wemelsfelder, F.; Fleming, P.A. Qualitative behavioural assessment of transport-naïve and transport-habituated sheep. *J. Anim. Sci.* **2012**, *90*, 4523–4535. [CrossRef] [PubMed]
6. Gutmann, A.K.; Schwed, B.; Tremetsberger, L.; Winckler, C. Intra-day variation of Qualitative Behaviour Assessment outcomes in dairy cattle. *Anim. Welf.* **2015**, *24*, 319–326. [CrossRef]
7. Minero, M.; Dalla Costa, E.; Dai, F.; Murray, L.A.M.; Canali, E.; Wemelsfelder, F. Use of Qualitative Behaviour Assessment as an indicator of welfare in donkeys. *Appl. Anim. Behav. Sci.* **2016**, *174*, 147–153. [CrossRef]
8. Kirchner, M.K.; Schulze Westerath, H.; Knierim, U.; Tessitore, E.; Cozzi, G.; Pfeiffer, C.; Winckler, C. Application of the Welfare Quality® assessment system on European beef bull farms. *Animal* **2014**, *8*, 827–835. [CrossRef] [PubMed]
9. Popescu, S.; Borda, C.; Diugan, E.A.; El Mahdy, C.; Spinu, M.; Sandru, C.D. Qualitative Behaviour Assessment of Dairy Cows Housed in Tie- and Free Stall Housing Systems. *Bull. UASVM Vet. Med.* **2014**, *71*, 276–277.
10. Temple, D.; Manteca, X.; Velarde, A.; Dalmau, A. Assessment of animal welfare through behavioural parameters in Iberian pigs in intensive and extensive conditions. *Appl. Anim. Behav. Sci.* **2011**, *131*, 29–39. [CrossRef]
11. Phythian, C.; Michalopoulou, E.; Duncan, J.; Wemelsfelder, F. Inter-observer reliability of Qualitative Behavioural Assessments of sheep. *Appl. Anim. Behav. Sci.* **2013**, *144*, 73–79. [CrossRef]
12. Welfare Quality®. *Welfare Quality® Assessment Protocol for Cattle*, corrected 1st Version; Welfare Quality® Consortium: Lelystad, The Netherlands, 2013.

13. Kirchner, M.K.; Schulze Westerath-Niklaus, H.; Gutmann, A.K.; Peiffer, C.; Tessitore, E.; Cozzi, G.; Knierim, U.; Winckler, C. Qualitative Behaviour Assessment is independent from other parameters used in the Welfare Quality® assessment system for beef cattle. In *ISAE2012—Quality of Life in Designed Environments*; Waiblinger, S., Winckler, C., Gutmann, A.K., Eds.; Wageningen Academic Publishers: Vienna, Austria, 2012; p. 79.
14. Tremetsberger, L. *Animal Health and Welfare Planning in Dairy Cattle—Effects on Animals and Farm Efficiency*; Department for Sustainable Agricultural Systems, Division of Livestock Science, University of Natural Resources and Life Sciences, Vienna (BOKU): Vienna, Austria, 2016; p. 82.
15. Gratzer, E.; Vasseur, E.; Winckler, C.; Schulze, W.H.; Piment, T.; Knierim, U.; van Reenen, K.; Buist, W.; Engel, B.; Lensink, J. *Final Report on a Prototype Welfare Assessment System for Dairy Calves and Rearing Heifers, the Final, Full Assessment System and on Risk Factor Analysis for Welfare Parameters in Dairy Calves and Rearing Heifers*; Welfare Quality Consortium: Lelystad, The Netherlands, 2010; p. 73.
16. Beaver, A.; Meagher, R.K.; von Keyserlingk, M.A.G.; Weary, D.M. Invited review: A systematic review of the effects of early separation on dairy cow and calf health. *J. Dairy Sci.* **2019**, *102*, 5765–5783. [CrossRef] [PubMed]
17. Ellingsen, K.; Coleman, G.J.; Lund, V.; Mejdell, C.M. Using qualitative behaviour assessment to explore the link between stockperson behaviour and dairy calf behaviour. *Appl. Anim. Behav. Sci.* **2014**, *153*, 10–17. [CrossRef]
18. Thomsen, P.T.; Sørensen, J.T. Factors affecting the risk of euthanasia for cows in Danish dairy herds. *Vet. Rec.* **2009**, *165*, 43–45. [CrossRef] [PubMed]
19. Torsein, M.; Jansson-Mörk, M.; Lindberg, A.; Hallén-Sandgren, C.; Berg, C. Associations between calf mortality during days 1 to 90 and herd-level cow and production variables in large Swedish dairy herds. *J. Dairy Sci.* **2014**, *97*, 6613–6621. [CrossRef]
20. Alvåsen, K.; Roth, A.; Jansson Mork, M.; Hallen Sandgren, C.; Thomsen, P.T.; Emanuelson, U. Farm characteristics related to on-farm cow mortality in dairy herds: A questionnaire study. *Animal* **2014**, *8*, 1735–1742. [CrossRef]
21. Burow, E.; Thomsen, P.T.; Sørensen, J.T.; Rousing, T. The effect of grazing on cow mortality in Danish dairy herds. *Prev. Vet. Med.* **2011**, *100*, 237–241. [CrossRef]
22. Zuliani, A.; Romanzin, A.; Corazzin, M.; Salvador, S.; Abrahantes, J.C.; Bovolenta, S. Welfare assessment in traditional mountain dairy farms, above and beyond resource-based measures. *Anim. Welf.* **2017**, *26*, 203–211. [CrossRef]
23. Simon, G.E.; Hoar, B.R.; Tucker, C.B. Assessing cow–calf welfare. Part 2, Risk factors for beef cow health and behavior and stockperson handling. *J. Anim. Sci.* **2016**, *94*, 3488–3500. [CrossRef]
24. R development Core Team. *R—A Language and Environment for Statistical Computing*; R development Core Team: Vienna, Austria, 2015.
25. Budaev, S.V. Using Principal Components and Factor Analysis in Animal Behaviour Research, Caveats and Guidelines. *Ethology* **2010**, *116*, 472–480. [CrossRef]
26. Jensen, M.B.; Vestergaard, K.S.; Krohn, C.C.; Munksgaard, L. Effect of single versus group housing and space allowance on responses of calves during open-field tests. *Appl. Anim. Behav. Sci.* **1997**, *54*, 109–121. [CrossRef]
27. Bøe, K.; Færevik, G. Grouping and social preferences in calves, heifers and cows. *Appl. Anim. Behav. Sci.* **2003**, *80*, 175–190. [CrossRef]
28. Bach, A.; Ahedo, J.; Ferrer, A. Optimizing weaning strategies of dairy replacement calves. *J. Dairy Sci.* **2010**, *93*, 413–419. [CrossRef] [PubMed]
29. Cobb, C.J.; Obeidat, B.S.; Sellers, M.D.; Pepper-Yowell, A.R.; Hanson, D.L.; Ballou, M.A. Improved performance and heightened neutrophil responses during the neonatal and weaning periods among outdoor group-housed Holstein calves. *J. Dairy Sci.* **2014**, *97*, 930–939. [CrossRef] [PubMed]
30. Muri, K.; Stubsjøen, S.M.; Valle, P.S. Development and testing of an on-farm welfare assessment protocol for dairy goats. *Anim. Welf.* **2013**, *22*, 385–400. [CrossRef]

 © 2019 by the authors. Licensee MDPI, Basel, Switzerland. This article is an open access article distributed under the terms and conditions of the Creative Commons Attribution (CC BY) license (http://creativecommons.org/licenses/by/4.0/).

Article

The Effects of Play Behavior, Feeding, and Time of Day on Salivary Concentrations of sIgA in Calves

Katrin Spiesberger [1,†], Stephanie Lürzel [1,*,†], Martina Patzl [2], Andreas Futschik [3] and Susanne Waiblinger [1]

[1] Institute of Animal Welfare Science, Department for Farm Animals and Veterinary Public Health, University of Veterinary Medicine, Vienna, Veterinärplatz 1, 1210 Vienna, Austria
[2] Institute of Immunology, Department for Pathobiology, University of Veterinary Medicine, Vienna, Veterinärplatz 1, 1210 Vienna, Austria
[3] Department of Applied Statistics, JK University Linz, Altenberger Str. 69, 4040 Linz, Austria
* Correspondence: stephanie.luerzel@vetmeduni.ac.at
† These authors contributed equally to this work.

Received: 26 May 2019; Accepted: 2 September 2019; Published: 5 September 2019

Simple Summary: The focus of animal welfare science has shifted over the last decades from efforts to avoid negative states to ways of allowing animals the experience of positive emotions. The emotional state of an animal interacts with its immune system. Secretory immunoglobulin A, a class of antibodies present on mucosal surfaces and acting as the first line of defense against infections, is influenced by positive and negative emotions in humans; the few studies of its association with emotions in animals focused almost exclusively on the impact of negative emotions and yielded conflicting results. We present the first study that focuses on salivary immunoglobulin A to investigate a possible relationship between positive emotions and immune functioning in calves. We detected a circadian rhythm of immunoglobulin A concentrations, with lowest levels at 14:00 h. Immunoglobulin A concentrations were decreased directly after feeding, possibly due to increased saliva flow rates, and we did not find higher immunoglobulin A concentrations after play. The results are important for the design of future studies of positive emotions, although they do not support immunoglobulin A as an indicator of positive emotional states.

Abstract: The focus of animal welfare science has shifted over the last decades from efforts to avoid negative states to ways of allowing animals the experience of positive emotions. They may influence physiological processes in farmed animals, potentially providing health benefits; in addition, the physiological changes might be used as indicators of emotional states. We investigated calves' salivary secretory immunoglobulin A (sIgA) concentrations with regard to a possible circadian rhythm and two situations that elicit positive emotions. Ten saliva samples of 14 calves were taken on two consecutive days; within the course of a day we observed a significant decline in salivary sIgA concentrations at 14:00 h. Further, we probed the animals before and after milk feeding and, contrarily to our prediction, detected lower sIgA concentrations 5 min after feeding than 15 min before. A probable explanation might be an increase in salivary flow rate caused by milk ingestion. We also took samples before and after we stimulated play behavior in calves. There was no significant difference in sIgA concentrations between samples taken before and after play. Although there was a significant correlation between the change in sIgA concentrations and the amount of play behavior shown, the correlation depended on an unexpected decrease of sIgA in animals that played little, and thus, does not support our hypothesis. In general, the data showed a large variability that might arise from different factors that are difficult to standardize in animals. Thus, the use of salivary sIgA concentrations as a marker of positive emotions in calves is not supported conclusively by the present data.

Keywords: immunoglobulin A; saliva; cattle; emotions; circadian rhythm

1. Introduction

Although the focus of animal welfare science has traditionally been mainly on negative aspects, it recently has shifted to include the assessment of positive welfare and thus, positive emotions [1]. In animal emotion research, the actual interest lies in the conscious subjective experience, characterized by arousal and valence [1,2]. Although this subjective component is not directly accessible to science, the corresponding behavioral, physiological, and cognitive components can be measured, making it possible to assess affective states in animals (e.g., [3]).

The influence of stress and affective states on immune functioning is well known, mostly in humans [4,5] but also in animals [6,7]. Measures of immune functioning have been proposed for assessment of affective states in animals [1]. One of the numerous indicators of immune functioning is immunoglobulin A (IgA) [8]. Secretory immunoglobulin A (sIgA) is present on most mucosal surfaces, providing the first line of defense of the organism against infective agents like bacteria and viruses [8]. In animal welfare studies, it has mainly been measured in saliva (e.g., [9]) and feces (e.g., [10]), but also in milk [11]. Salivary sIgA concentrations were shown to react within 10–15 min after eliciting an emotional state in animals [9,12].

In general, data on sIgA concentrations after the experience of either positive or negative emotions in animals are limited (for a review, see [8]). The few published studies focused mostly on negative emotions, with conflicting results. In dogs, decreased sIgA concentrations were reported after stressful situations [9,13]; the effect appears to be age-dependent, since sIgA concentrations were increased in puppies after stress [14]. In male rats, social housing conditions with different levels of competition and mating opportunities influenced salivary sIgA concentrations, with conditions deemed more favorable (presence of a female and bedding) leading to a steep decrease and subsequently to a gradual increase and less favorable conditions (group-housing with other males) to a decrease [15]. In pigs, an increase in salivary sIgA was detected after restraint stress [12]. Regarding cattle, there is one study that examined the effect of removal of conspecifics on sIgA [11], which is a stressful event according to behavioral observations. After the removal, they found no difference in milk sIgA concentrations between cows that were associated with the removed animals or not; there was an increase in serum IgA concentrations in calves and young bulls after removal of their pen mates, but sample sizes were very small and there were no control groups [11].

Although there are strong indications for an increase of salivary sIgA in response to positive emotions in humans (e.g., [16,17]), there are few studies in animals. Regarding a potential effect of positive emotions, sIgA concentrations in the feces of sheltered cats that had been stroked or whose behaviors were classified as positive were elevated [10,18]. To be able to interpret results on sIgA correctly, the circadian rhythm should be considered. The circadian rhythm may vary strongly between species [8] and although there are studies in dogs [9] and pigs [12], no data are available for cattle to date. The first aim of our study was therefore to investigate a potential circadian rhythm of salivary sIgA dynamics in calves. The second and main aim was to elucidate the effect of positively valenced emotions on salivary sIgA in calves. We hypothesized an increase of sIgA during milk feeding as well as during experimentally induced play behavior, which are both situations that are associated with positive emotions [1].

2. Materials and Methods

2.1. Subjects, Housing and Management

Twelve Austrian Simmental (eight females, four males) and two Holstein calves (both female) were tested for changes in salivary sIgA concentrations during play at an average age of 61 ± 9 days (mean ± SD). From 14 days of age, the calves were housed together in a deep litter barn at the Teaching and Research Estate Kremesberg of the Vetmeduni Vienna (Pottenstein, Austria). They were kept in three groups of six to eight animals together with calves that were not involved in the study. We tested two animals out of six in group A, eight animals out of eight in group B and four out of seven in group C. Each group was housed in a 7 × 5 m deep litter pen, including a 12.5 m² area of 1.3 m height that was separated by a transparent strip curtain, and eight individual feeding stalls (0.5 × 1.7 m²) with wooden walls and concrete floor. The calves were fed with pasteurized milk twice a day, around 07:30 h and 18:00 h, with 3–5 L of milk per feeding, depending on age. During milk feeding until approximately 30 min after feeding, the calves were confined in the feeding stalls by gates to reduce allosucking; during the rest of the day, they could enter and leave the stalls freely. Water, calf feed (Kälber Start Vital; Garant, Pöchlarn, Austria) and hay were provided ad libitum. All calves were disbudded with a hot iron by a veterinarian at an age of 4–5 weeks. They were sedated (Sedaxylan, 20 mg Xylazin/mL: 0.1 mg Xylazin/kg body weight) and received local anesthesia (Procamidor 2% Procain: each side ca. 5 mL) and analgesia (Rifen, 100 mg Ketoprofen/mL: 3 mg Ketoprofen/kg body weight). Disbudding was performed at least 14 days before the habituation period started. The study was discussed and approved by the institutional ethics committee in accordance with Good Scientific Practice (GSP) guidelines [19] and national legislation (project number 01/03/97/2014).

2.2. Experimental Procedures

The experiment took place between March and May 2014 (Figure 1). After the habituation period, the salivary sIgA concentrations of the subjects were determined over the course of the day on two consecutive days per calf in order to determine the circadian sIgA rhythm and the possible influence of milk feeding. Between 4 and 14 days later (depending on temporal constraints due to farm procedures), saliva samples were taken before and after induced play and behavior was directly observed as well as video-recorded for later, detailed analysis.

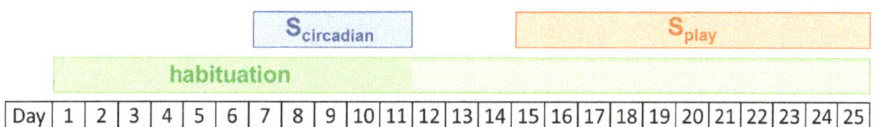

Figure 1. Schematic view of experimental phases. After the first days of habituation, saliva was sampled for detecting a circadian rhythm ($S_{circadian}$) on two consecutive days for each calf (between day 7 and day 11). Saliva sampling before and after play (S_{play}) took place on days 15 to 25. During the first 11 days, the habituation procedures were performed daily (dark green). To maintain familiarity of the calves with the experimenters, habituation was continued after $S_{circadian}$ on every second day (light green).

Calves were habituated to human contact and the procedure of saliva sampling by one of two female experimenters (170 cm, blonde, and 176 cm, brown hair and glasses, both dressed in dark green overalls) for 1–2 h twice a day for 11 consecutive days (Figure 1). First, the experimenter entered the box and initially waited for the calves to seek contact. As soon as the animals approached the experimenter voluntarily and did not show overt avoidance reactions, she also initiated body contact, stroked them, and allowed them to suck her fingers. If the calf pulled away after initial contact, the experimenter waited for the calf to approach again. Furthermore, the procedure of saliva sampling was simulated by placing the sampling device shortly into the calves' mouths. The experimenters did not encourage the subjects to play nor actively played with them during the habituation period. Calves were considered as sufficiently habituated when they approached the experimenter readily and/or let themselves be touched. After taking the samples for the analysis of the circadian rhythm, the procedure of habituation was continued every second day for 14 more days (day 11 to day 25) until all the samples in the context of the play situation were taken (Figure 1).

On two consecutive days, saliva samples were taken to gain general information about each calf's salivary sIgA concentration during the course of a day and the influence of feeding. Depending on the calves' state of habituation, they were sampled after 7–11 days of habituation, and on any given day, two to four calves were sampled. Six samples were taken over the course of each day (Figure 2), at 07:15 h (before morning milk feeding), 10:00 h, 12:00 h, 14:00 h, 16:00 h, and 17:45 h (before evening milk feeding). The samples at 07:15 h and 17:45 h were also used as a baseline in the analysis of a possible effect of feeding on sIgA concentrations. Additional samples were taken 5 and 30 min after the end of milk feeding in the morning and in the evening, resulting in a total of 280 saliva samples. The removal of the feeding bucket as soon as it was empty marked the end of milk feeding, which varied among the tested calves, depending on the amount of milk fed and each calf's drinking behavior. As a circadian rhythm with a significant change within 35 min is highly improbable, we refrained from including a control condition for feeding; a "no feeding" condition would not have been not valid control, as the animals would have anticipated to be fed at this time of day, and thus, there might have been effects of emotional state (frustration), saliva flow and possibly other factors on sIgA concentrations.

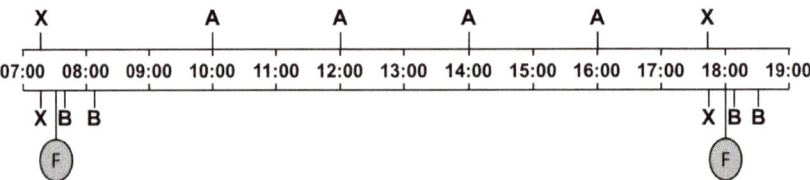

Figure 2. Schedule for saliva sampling in calves. F, morning and evening milk feeding (07:30 h and 18:00 h); A, saliva sampling to determine circadian rhythm at fixed times of day; B, saliva sampling to determine effect of feeding, 5 and 30 min after the end of milk feeding; X, saliva sampling serving both purposes.

Play tests were performed 4 to 14 days later, twice a day, around 10:15 h as well as 12:30 h, i.e., late enough to exclude a possible influence of feeding. Every calf was tested maximally two times per day and in up to four tests in total. Play behavior cannot be triggered reliably: Sometimes some calves are not motivated to play, whereas at other times, several calves will play at the same time. In addition, play behavior is often contagious [20], i.e., if one calf starts playing, others will join in [21,22]. Thus, we aimed to exploit the effects of emotional contagion or social facilitation and tested several calves in the same pen at the same time. Usually two to four calves were tested per play session depending on the number of experimental animals kept in the pen. For the test, an experimenter entered the box and immediately took a baseline saliva sample of the subjects to be tested. Then locomotor play behavior was encouraged by the experimenter moving between the calves. The way of moving included running and jumping with relatively slow, exaggerated movements, sometimes but not always including eye contact with the calves. Further, the experimenter initiated physical contact with the calves to induce play fighting behavior: She touched or rubbed a calf's forehead and progressed to pushing if the animal started to perform head pushing or rubbing. In the meanwhile, the other experimenter manually recorded the frequency of play behavior shown by the subjects to be tested. Depending on whether and when such behavior was shown—according to the second experimenter's observations—a second saliva sample was taken 15–45 min after the baseline sample; the experimenter aimed to take it within 8 min after play behavior was shown [23].

Saliva was sampled using Salivettes® (Sarstedt; Nürnbrecht, Germany). A rubber teat from a bucket was put over dressing forceps so that approximately 2 cm of the forceps were visible outside of the rubber teat. A cotton swab was then gripped with the forceps. For taking a saliva sample, the experimenter approached a calf in its home pen and carefully placed the forceps into its mouth for at least 30 s, if possible without restraining the subject. Most often, simultaneous stroking of the calf was sufficient to make it tolerate the sampling procedure; only for 29 out of 334 samples, the subject had to be restrained by the experimenter. If restraint was necessary, the calf was either confined in a feeding stall or held by the experimenter by putting one arm around its neck. All samples that were used to analyze the effect of feeding on IgA were taken while the animals were in the feeding stalls. Samples were immediately put on ice and frozen at −20 °C within a maximum of 15 min.

2.3. Analysis of sIgA Concentrations

Saliva samples were thawed and centrifuged for 5 min (1000× g, 4 °C). Supernatant was taken and samples were diluted in Tween-TRIS buffer (50 mM Tris, 0.14 M NaCl, 0.05% Tween 20, pH 8) at ratios of 1:1000, 1:2500, 1:5000, and 1:10,000. Salivary sIgA concentrations were determined using the Bovine IgA ELISA Quantitation Set according to the manufacturer's protocol (Bethyl Laboratories; Montgomery, AL, USA). Standard and sample dilutions were analyzed in duplicates. Optical densities (OD) at 450 nm were measured using an ELISA reader (Epoch Microplate Spectrophotometer, Biotek; Bad Friedrichshall, Germany). The OD values and log IgA concentrations were plotted using a four-parameter logistic regression model analysis.

2.4. Observation of Play Behavior

Behavior was observed directly and coded from video recordings after training by one of the experimenters, using an ethogram (Table 1) based on previous descriptions [21,22,24]. It included social and locomotor play behavior patterns as well as avoidance, because the occurrence of repeated avoidance may indicate a certain level of fear that could influence salivary sIgA concentrations. Head rubbing towards objects in the environment has been described in the context of play behavior [24]. As we did not expect the experimenter to induce object play in the calves, we focused on social play and included social head rubbing because it is often shown in the context of play behavior and there are gradual transitions between head rubbing and head pushing. In addition to this playful component, it is also an affiliative behavior [22] and thus highly likely to contribute to or indicate a positive affective state [1], which we intended to induce. Direct observations were necessary to determine when the second salivary sample should be taken, whereas video observations allowed coding play and avoidance behavior in detail. Behavior was observed directly by an experimenter sitting in front of the tested calves' pen at about 2 m height in the feeding alley. For video recordings, two cameras (EcoLine TV7204, Abus; Wiener Neudorf, Austria) were placed at 3 m height, one at the left and one at the right corner of the front side of the pens. The observation started as soon as the experimenter entered the pen and ended when she left it. Only the behavior that was shown between the two saliva samples was analyzed. Behavior was coded using the Interact® software (V14.0, Mangold; Arnstorf, Germany), recording frequencies and durations of behavioral patterns (Table 1). The observer also recorded when saliva samples were taken. Both types of behavioral observations were done using behavior sampling and continuous recording [25].

Table 1. Ethogram of observed behaviors. For running, head pushing and head rubbing, durations were recorded; all other behaviors were recorded as frequencies. Head shaking is a supplemental definition (to distinguish avoidance from playful running that is directed away from the experimenter) and was not recorded on its own, as it usually occurs together with running or jumping. All behaviors except for avoidance were considered play behaviors (described in [21,22,24]).

Behavior	Definition
Running	A calf moves forward faster than walking (trot or gallop).
Jumping	During running: All four legs leave the ground, accompanied by a clear upward movement of the calf. On the highest point of movement, the animal can kick with one or both hind limbs. During standing/walking: In an upward movement, both forelimbs leave the ground and the calf lands with both forelimbs simultaneously. The hind limbs can also move.
Kicking	While standing or walking, the calf kicks with one hind limb.
Mounting	A calf jumps with both forelimbs and lays the front part of its body on the body of another animal or the experimenter. Mounting is also recorded if the attempt is not successful, i.e., the upper body part does not come to rest on the other animal or the experimenter.
Head pushing	A calf puts its forehead against the forehead or head/neck region of another calf or against a body part of the experimenter and pushes. This behavior can also be started with another part of the head than the forehead.
Head rubbing	A calf puts any part of its head, usually the side of the face, against a body part of another calf and rubs it in an up-and-down movement.
Head shaking	Up-and-down or rotational head movements, often in combination and around more than one axis; the movements have no clear direction, e.g., towards flies.
Avoidance	A calf moves away from the experimenter after the experimenter moved towards the calf. This behavior is only recorded if the movement is obviously triggered by the experimenter's approach. If this experimenter-triggered, averted movement leads to a clearly playful behavior (jumping, kicking, head shaking), it is not recorded as avoidance but as the according play behavior (in case of a quick movement in combination with head shaking, the behavior is recorded as running).

2.5. Statistical Analysis

All statistical analyses were performed and graphs were created using the R statistical environment, versions 3.2.3 and 3.4.3 [26]. In all boxplots, the length of the box refers to the interquartile range (IQR) and the horizontal line represents the median value. The end of the lower whisker represents the lowest data point still within 1.5 × IQR from the lower quartile and the end of the higher whisker represents the highest data point still within the 1.5 × IQR from the upper quartile. Values outside this range are considered outliers and depicted as open dots.

The experimental unit was the individual calf. To analyze the data gained from saliva samples taken over the course of the day (2 × 10 samples per calf), linear mixed models (LMM) were calculated. In all models, the animal nested in the group was included as a random effect. For the analysis of changes in circadian sIgA concentrations, time of sampling, sex, day of sampling and their interactions were included as fixed effects in the full model. After model selection using the Akaike Information Criterion (AIC), only time was included as a fixed effect. For the analysis of a possible influence of feeding on IgA concentrations, sample number (SB, baseline sample taken before feeding; S5, taken 5 min after feeding; S30, taken 30 min after feeding), time of day (AM; PM), sex, day of sampling and their interactions were included as fixed effects in the full model. After model selection using the AIC, only sample number was included as a fixed effect. Normal distribution of residuals was checked by means of a Q-Q plot, the homogeneity of variance of the residuals by plotting the residuals against the predicted values resulting from the model. sIgA values were log-transformed to fulfill the model assumptions. If there were outliers in the residuals (>3 SD), the model was calculated again without the corresponding observations in order to determine whether they have a strong influence on the results. This was never the case and the models calculated before the exclusion of outliers are presented. Post-hoc comparisons were calculated with Tukey adjustment.

One calf was excluded from analysis of play behavior and its effect on sIgA due to loss of video data. To summarize play behavior, a duration of 1 s was assigned to each occurrence of behaviors recorded as events, because their duration would usually be around 1 s as observed during the habituation phase. A play score was then calculated for each play test of each calf by adding the durations of all play behaviors, divided by the time (in min) between the two saliva samples to account for different durations of observation. If several play tests were conducted with a calf, only the test with the highest play score was chosen for the analysis of salivary sIgA concentrations. If a calf was tested only once, the samples were included in the analysis regardless of the calf's play score. A one-sample sign test was performed to compare salivary sIgA concentrations before and after playing. Since the average duration of a play test was relatively short (mean 24.9 ± SD 7.5 min), sIgA concentrations were not expected to be influenced by the circadian rhythm, and we refrained from controlling for an effect of time. Delta sIgA was calculated (ΔsIgA = sIgA concentrations determined after the play test minus baseline sIgA concentrations determined before the play test) and a possible correlation of the play score and ΔsIgA was investigated using Spearman's rank correlation test, as it is insensitive against outliers and we did not expect a strictly linear correlation. The same test was used for calculating the correlation between play score and avoidance behavior.

3. Results

3.1. Effect of Time of Day

The mean salivary sIgA concentration was 689 ± 1115 µg/mL (mean ± SD) on the first day and 636 ± 803 µg/mL on the second day. Three outliers with sIgA concentrations exceeding 3000 µg/mL were observed on the first day and one outlier on the second day (Table S1). There was a significant effect of time (Figure 3, LMM, $F_{5,149} = 5.4$, $p < 0.001$), with lower sIgA concentrations at 14:00 h than at all other points in time ($p = 0.001$–0.022).

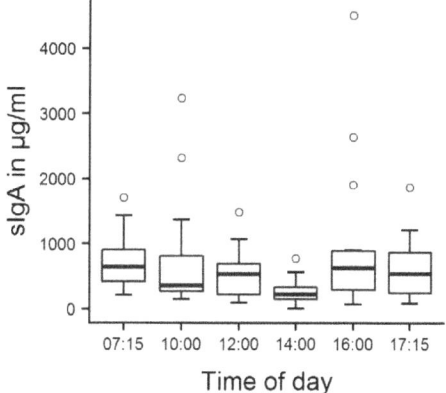

Figure 3. Calves' salivary sIgA concentrations at different points in time. Fourteen calves were tested on two consecutive days; values were averaged per calf and point in time. There was a main effect of time (LMM, $p < 0.001$); the values at 14:00 were significantly lower than the values at all other points in time according to Tukey contrasts ($p = 0.001$–0.022).

3.2. Effect of Feeding

The mean sIgA concentration was 533 ± 504 µg/mL on the first day and 555 ± 531 µg/mL on the second day. There was a main effect of sample number on salivary sIgA concentrations (Figure 4, LMM, F_{2152} = 5.2, p = 0.007): salivary sIgA concentrations were lowest 5 min after milk feeding (Table S2; p = 0.007) and increased afterwards (p = 0.049).

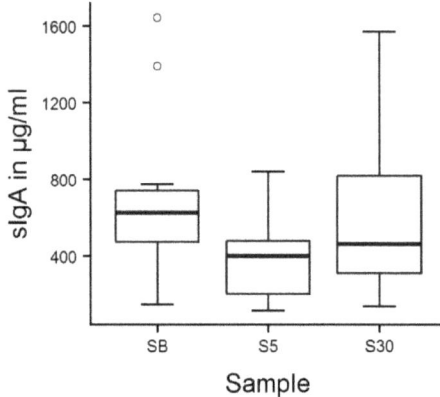

Figure 4. Calves' salivary sIgA concentrations around feeding time. Fourteen calves were sampled 15 min before (SB), 5 min (S5), as well as 30 min after the milk feeding (S30), which took place twice a day on two consecutive days; values were averaged per calf and point in time. There was a main effect of sample number (LMM, p = 0.007); values decreased from SB to S5 (p = 0.007) and increased from S5 to S30 (p = 0.049) according to Tukey contrasts.

3.3. Play Behavior and Its Effect on sIgA Concentration

Within the play test, it took the experimenter on average 3.0 ± 2.5 min to take the baseline saliva sample after she entered the pen. Out of 13 calves, two showed short occurrences (<30 s) of play behavior in the time from entering the pen until the experimenter took the sample. On average, the calves showed their first play behavior 5.8 ± 2.9 min after the baseline saliva sample was taken. The second saliva sample was taken 24.9 ± 7.5 min after the baseline sample. Between the two samplings, the experimenter induced play behavior (Figure 5; Table S3). On average, 18.9 ± 7.9 min elapsed between the calves' first play behavior and the collection of the second saliva sample and 3.1 ± 3.3 min between the last play behavior and the second sample.

The play score ranged from 0.5 to 5.6, with a mean score of 2.6 ± 1.7. In total, avoidance behavior was recorded 37 times (0.14 ± 0.18 times/min), but it was mostly shown by only two calves (play scores: 1.6 and 3.8). The calves that played most (play scores of 5.6 and 5.1) never avoided the experimenter in the play test. However, there was no significant correlation between avoidance behavior and play score (Spearman's rank correlation, r = −0.28, p = 0.35). The salivary sIgA concentrations of the calves probed before the play test ranged from 13 to 1030 µg/mL (mean: 283 ± 265 µg/mL). After the play test, sIgA concentrations were between 22 and 512 µg/mL (mean: 258 ± 143 µg/mL). There was no significant difference between sIgA concentrations before and after the play test (Figure 6, one-sample sign test, s = 7, p = 0.39) but we found a strong, significant correlation between the play score and ΔsIgA (Figure 7, Spearman's rank correlation, r = 0.69, p = 0.012).

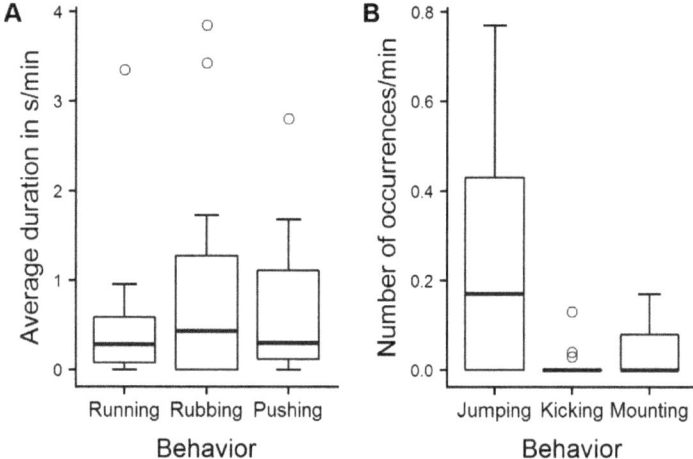

Figure 5. Play behavior. Average duration in s/min (**A**) and number of occurrences/min (**B**) of different patterns of play behavior displayed by 13 calves in the play test selected for evaluation of the effect of play behavior on sIgA.

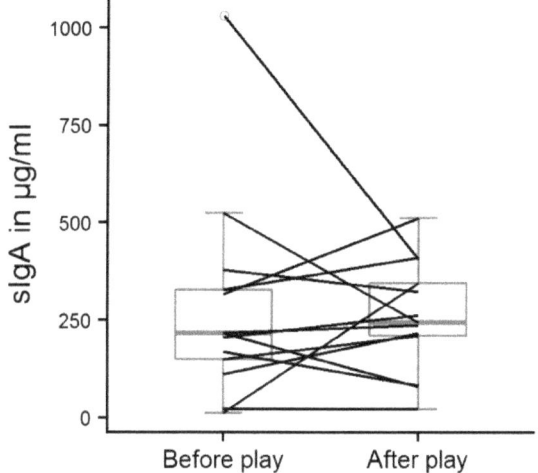

Figure 6. Salivary sIgA concentrations before and after the play test situation. One-sample sign test, n = 13, s = 7, $p = 0.39$.

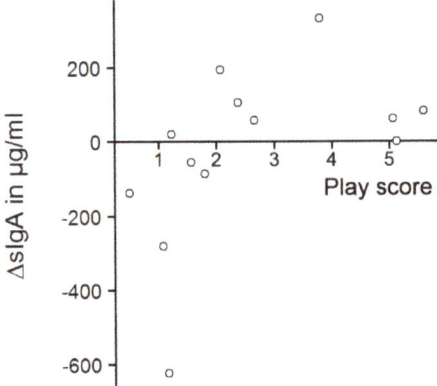

Figure 7. ΔsIgA and play score. ΔsIgA = (sIgA concentration after play test) minus (sIgA concentration before play test); Spearman's rank correlation, n = 13, r = 0.69, p = 0.012.

4. Discussion

4.1. Effect of Time of Day

There was an effect of time on calves' salivary sIgA concentrations, with the lowest concentrations detected at 14:00 h, compared to all other times; sIgA seems thus to undergo circadian changes. In humans, salivary sIgA was shown to decline steadily during the course of the day [27], but little is known about animals' circadian changes of sIgA concentrations [8]. Contrary to our findings, studies in pigs and dogs demonstrated lower salivary sIgA in the morning and elevated salivary IgA concentrations over the midday hours, which declined again in the late afternoon [9,12]; rats housed in metabolic cages excreted the highest amounts of fecal sIgA between 20:00 h and 02:00 h, which corresponds to a major part of their active phase [28]. The different results might be due to general differences between species or their different diurnal rhythms of activity or arousal. In calves, rumination might also influence salivary sIgA concentrations, since it may result in enhanced saliva flow rates that lead to a decline in sIgA concentration [29], and additionally in mixing of rumen content with saliva.

In general, the data showed a high variability between and within the calves as well as large outliers. From a technical point of view, the large variability of our data could be influenced by the restricted possibilities of standardization of the sampling procedures in animals. In humans, saliva sampling is a well-established procedure that includes a standardized behavioral protocol, resulting in saliva collected over a defined time span and under comparable conditions [30]. Saliva sampling is far more difficult to standardize in animals because they cannot be instructed to behave in the way a human can be. Consequently, the time span of sampling differs, as some animals interrupt the sampling. Also, oral behavior before the sampling procedure could not be controlled in the present study, which is important since salivary sIgA is dependent on the saliva flow rate, which is influenced by mechanical, gustatory and olfactory factors [31].

Salivary sIgA concentrations can change in response to positive or negative emotional events and salivary IgA concentrations react to emotional stimulation already several minutes after the onset of stimulation [9,32]. Since we cannot exclude that tested calves experienced positive or negative emotions before samples were taken, emotional experiences constitute a potential confounder. In future studies, behavior should be observed for several hours before sampling to detect such situations and consider them in data analysis. However, it is unlikely that calves experience strong emotions at consistent times across days, outside of the context of milk feeding.

4.2. Effect of Feeding

The examination of salivary sIgA concentrations revealed decreased concentrations directly after feeding, regardless of the tested day, time of feeding or sex of calves. There was no significant difference between the samples taken before and 30 min after the milk feeding. As we rated feeding as a positively valenced event, we would have expected an increase in salivary sIgA concentrations directly after feeding, which was clearly not the case. Food intake stimulates saliva production [31], and with increasing saliva flow rates, salivary sIgA concentration decreased in humans [29]. Thus we suppose that the observed decline in sIgA concentrations resulted from an increase of saliva flow rate during feeding. As it was not possible to determine saliva flow rate in our experiment, there are no data to support this explanation at the moment. Emotional arousal is a component of emotional states that can act as a confounder. As we had no possibility to control for potential differences in arousal in the present experiment, the observed sIgA decline might theoretically also be explained by a state of low arousal caused by post-consummatory satisfaction [2].

Another factor that could have influenced the obtained results is the temporal dynamics of sIgA in saliva. Little information is available about temporal aspects of salivary sIgA, but concentrations were significantly different to baseline at 10 and 15 min after the onset of an emotional stimulus in dogs and pigs, respectively [9,12]. These samples were, however, the first samples taken after the onset of the stimulus, which means that sIgA might react even faster. Pre-stimulus levels were reached 10 min (pigs) and 30 min (dogs) after the end of the stimulus, although still being significantly different from controls in dogs. Again, this was the first recovery sample, so pre-stimulus levels might be attained even faster. As we took the first post-stimulus saliva sample approximately 10 min after the onset of feeding and 5 min after the bucket had been removed, it seems reasonable to assume that we indeed detected the sIgA response to feeding. Nevertheless, there might be differences between species, thus further research is necessary to determine the time course of the salivary sIgA reaction in cattle.

In a follow-up study, a possible sampling effect on our data should be considered; without having a non-feeding control it is not possible to distinguish the feeding effect from a potential sampling effect. Although the calves were habituated to all experimenters and the sampling procedure, we cannot exclude that sampling could have caused a negative affective state or stress in some animals that influenced the results, as discussed in detail in Section 4.3.

4.3. Effect of Play Behavior

We did not find any significant differences in calves' salivary sIgA concentrations between the samples taken before and after the play test. We had assumed that every playing calf experienced positive emotions with a moderate to high level of arousal and predicted an increase in sIgA concentrations, but sIgA concentrations were not generally increased after play behavior was performed, as it increased numerically only in seven of the thirteen tested calves. There was a correlation between play score and change in salivary sIgA concentrations, but the result should be interpreted with caution: the majority of the calves that had played a lot showed numerically higher sIgA concentrations, but the calves that played most showed no or weak elevation of sIgA concentrations. In addition, some of the calves that had shown only little play behavior actually had numerically lower sIgA concentrations at the second sample point, especially the ones with very high starting values. The correlation depends thus strongly on the unexpected decreases, and therefore does not support our hypothesis of increased sIgA values after a positive emotional state.

A possible explanation for the numerical decrease might be that the sampling of saliva caused a negative affective state or stress, which might have decreased sIgA concentrations directly or influenced it via an effect on saliva production. Conversely, this potential negative affective state might have masked the effect of play behavior on sIgA concentrations: in animals that showed little play behavior, the effect of play behavior might not have been strong enough to make up for the initial decrease. It is also possible that the animals that played a lot did not experience negative affect or stress during the sampling, which in turn might have facilitated the stimulation of play behavior. As we did not expect negative effects of the saliva sampling because the animals were usually not restrained during the procedure, there was no situation without play that could have been used to control for the effect of sampling itself. These possible explanations remain thus speculative until a follow-up study will be conducted, including a control for the effect of sampling.

In addition, the relationship between sIgA concentrations and play behavior might be complex. For instance, physical activity could have an influence on salivary sIgA: increased [33] as well as decreased [34] salivary IgA concentrations in humans were found during a period of intense exercise, but to our knowledge, no data on the influence of physical activity on sIgA in animals are available. In the present study, it was not possible to separate the influence of emotions from the influence of physical activity, as play behavior is inextricably linked with physical activity, and most physical activity occurred in the context of play behavior. Another confounding effect might have been anticipation, which might have increased basal sIgA concentration. An indication for anticipation was the behavior of the two calves that started to play when the experimenter entered the pen, even before she started to induce play behavior in the calves. However, it speaks against a confounding effect of anticipation that the calves with the highest basal sIgA concentrations played relatively little (play scores 1.16 and 1.07); if their sIgA concentrations were that high because of a premature rise due to anticipation, we would have expected them to show high levels of play behavior in the test.

Play behavior is not easy to define and quantify in many animal species, as it combines behaviors from different behavioral categories. One might thus argue that the behaviors we recorded in this study were possibly caused by fear of the calves towards humans. However, play behavior in calves encompasses quite specific behaviors, which have been described in detail [21,22,24]. The animals did not show defense behavior, which would have been expressed as aggressive behavior towards the experimenter and might have led to injury. Regarding fear, it is more difficult to distinguish, as running may be caused by fear or occur in the context of play behavior. We took a conservative approach and recorded every movement away from the experimenter that did not include specific behavioral elements clearly indicating play (such as kicking, head shaking) as "avoidance", if it occurred as a direct reaction to her approach. We can thus be quite sure that we recorded all avoidance events as such and rather under- than overestimated the occurrence of play behavior. Another aspect of this issue is that humans are very adept at picking up subtle, qualitative cues in an animal's behavior that allow them to draw conclusions about the animal's emotional state [35,36]; we are confident that we would have recognized fearful behavior.

Experiments involving human contact are always influenced by the animal-human relationship (AHR) [37,38]. A positive AHR was essential for conducting this study, as the experimenters depended on the calves' cooperation during saliva sampling and play behavior occurs usually only in a relaxed situation [20]. The play experiments were performed by two female experimenters, who were familiar to the animals. Nevertheless the experimenters' behaviors, which triggered play behavior in some calves, might have induced fear in others. Negative feelings towards the experimenter or the play situation should have been reflected by increased avoidance behavior and the absence of play behavior [20,38]. Although we cannot completely rule out that play behavior might have served as a coping mechanism [39,40], in this case to cope with potential fear towards the experimenter, we consider it highly unlikely in the present study: Although Ahloy-Dallaire et al. [39] present several counter-examples to the wide-spread view that play is associated with positive affect, the type of play behavior shown in the described situations seems to be suitable to actually improve the situation and provide a welfare or fitness benefit, such as social play in socially stressed primates or object-play in nutritionally deprived kittens. In contrast, calves' primary response towards a truly threatening stimulus should be flight and not play, if we think in terms of evolutionary adaptiveness. Avoidance behavior occurred several times; it was mostly shown by only two calves with low and intermediate play scores. On the other hand, some calves were apparently waiting for the experimenter to enter the pen, immediately seeking body contact and sometimes initiating play behavior, showing no or minimal avoidance behavior. The AHR may have differed in our experimental animals, affecting their emotional state during the interactions and thus possibly their sIgA concentrations.

When it comes to emotions and sIgA, the available literature presents rather conflicting findings. Most of the experiments published on the connection between salivary sIgA and emotions were conducted in humans. There are several studies that indicate that positive emotions lead to increased salivary sIgA concentrations, e.g., [16,17], whereas stress as experienced during a period of academic exams down-regulated sIgA secretion [41]. However, Benham et al. [23] showed increased sIgA concentrations after stress-inducing as well as stress-reducing tasks; it is likely that the intensity and duration of the stressor play a role. Studies in animals neither agree on whether sIgA increases or decreases after positively experienced emotions. Authors reported decreased sIgA concentrations after stress-related situations in adult dogs and rats [9,13–15]. In contrast, increased salivary IgA concentrations after stressful situations were obtained in pigs and puppies [12,14]. The influence of positive emotions on salivary sIgA was only described in a small study of cattle: A trend towards increased salivary sIgA concentrations was observed after cows were moved to pasture after a long period of loose housing [42]. As in our study, a high variability in sIgA concentrations was reported, suggested to arise from environmental factors, and an unexpected, slight decrease was found mainly in animals with higher starting values.

The reported diversity between and within the species in salivary sIgA responses to differentially valenced emotions raises the question of the role of arousal in sIgA secretion. According to Mendl et al. [2], situations described as acute stress can be considered as experiences associated with high arousal whereas chronic stress, due to a habitation effect, might be connected to low arousal. To our knowledge, only Guhad and Hau [15] specifically investigated chronic stress in rats and observed decreased sIgA concentrations. All other studies dealt with experimental situations causing acute stress and described salivary IgA increases as well as decreases in highly aroused animals [9,12,14].

5. Conclusions

In this study, we detected a circadian rhythm of salivary sIgA concentrations in calves, with lowest concentrations at 14:00 h as well as reduced sIgA concentrations directly after feeding. There was no consistent response in sIgA after play, with both increasing and decreasing concentrations. The large variability of the data and the conflicting results in other mammalian species suggest additional influencing factors such as the sampling procedure, salivary flow rates, affective arousal or the AHR, as well as the current infectious pressure. Our results thus do not support the use of salivary sIgA concentrations as a marker of positive emotions in calves.

Supplementary Materials: The following are available online at http://www.mdpi.com/2076-2615/9/9/657/s1, Table S1: IgA concentrations over the course of the day, Table S2: IgA concentrations before and after feeding, Table S3: IgA concentrations and play behavior during the play tests.

Author Contributions: Conceptualization, S.L. and S.W.; methodology, S.L., S.W., A.F. and M.P.; formal analysis, K.S., S.L. and A.F.; investigation, K.S.; resources, M.P.; data curation, K.S. and S.L.; writing—original draft preparation, K.S. and S.L.; writing—review and editing, K.S., S.L., A.F., M.P. and S.W.; visualization, K.S. and S.L.; supervision, S.W.; project administration, S.L.

Funding: This research received no external funding.

Acknowledgments: We are grateful for the good cooperation with the management and staff of the Teaching and Research Estate Kremesberg (now VetFarm) of the University of Veterinary Medicine Vienna. We would also like to thank Barbara Mikolka and Christina Ederer for their help during data collection and two anonymous reviewers for their comments and suggestions. Open Access Funding was provided by the University of Veterinary Medicine, Vienna.

Conflicts of Interest: The authors declare no conflict of interest. The funders had no role in the design of the study; in the collection, analyses, or interpretation of data; in the writing of the manuscript, or in the decision to publish the results.

References

1. Boissy, A.; Manteuffel, G.; Jensen, M.B.; Moe, R.O.; Spruijt, B.; Keeling, L.J.; Winckler, C.; Forkman, B.; Dimitrov, I.; Langbein, J.; et al. Assessment of positive emotions in animals to improve their welfare. *Physiol. Behav.* **2007**, *92*, 375–397. [CrossRef]
2. Mendl, M.; Burman, O.H.; Paul, E.S. An integrative and functional framework for the study of animal emotion and mood. *Proc. Biol. Sci.* **2010**, *277*, 2895–2904. [CrossRef]
3. Mendl, M.; Burman, O.H.P.; Parker, R.M.A.; Paul, E.S. Cognitive bias as an indicator of animal emotion and welfare: Emerging evidence and underlying mechanisms. *Appl. Anim. Behav. Sci.* **2009**, *118*, 161–181. [CrossRef]
4. Muehsam, D.; Lutgendorf, S.; Mills, P.J.; Rickhi, B.; Chevalier, G.; Bat, N.; Chopra, D.; Gurfein, B. The embodied mind: A review on functional genomic and neurological correlates of mind-body therapies. *Neurosci. Biobehav. Rev.* **2017**, *73*, 165–181. [CrossRef]
5. Epel, E.S.; Crosswell, A.D.; Mayer, S.E.; Prather, A.A.; Slavich, G.M.; Puterman, E.; Mendes, W.B. More than a feeling: A unified view of stress measurement for population science. *Front. Neuroendocrinol.* **2018**, *49*, 146–169. [CrossRef]
6. Chebel, R.C.; Silva, P.R.B.; Endres, M.I.; Ballou, M.A.; Luchterhand, K.L. Social stressors and their effects on immunity and health of periparturient dairy cows. *J. Dairy Sci.* **2016**, *99*, 3217–3228. [CrossRef]
7. Gimsa, U.; Tuchscherer, M.; Kanitz, E. Psychosocial stress and immunity—what can we learn from pig studies? *Front. Behav. Neurosci.* **2018**, *12*, 64. [CrossRef]
8. Staley, M.; Conners, M.G.; Hall, K.; Miller, L.J. Linking stress and immunity: Immunoglobulin A as a non-invasive physiological biomarker in animal welfare studies. *Horm. Behav.* **2018**, *102*, 55–68. [CrossRef]
9. Kikkawa, A.; Uchida, Y.; Nakade, T.; Taguchi, K. Salivary secretory IgA concentrations in beagle dogs. *J. Vet. Med. Sci.* **2003**, *65*, 689–693. [CrossRef]
10. Gourkow, N.; Hamon, S.C.; Phillips, C.J. Effect of gentle stroking and vocalization on behaviour, mucosal immunity and upper respiratory disease in anxious shelter cats. *Prev. Vet. Med.* **2014**, *117*, 266–275. [CrossRef]
11. Walker, J.K.; Arney, D.R.; Waran, N.K.; Handel, I.G.; Phillips, C.J.C. The effect of conspecific removal on behavioral and physiological responses of dairy cattle. *J. Dairy Sci.* **2015**, *98*, 8610–8622. [CrossRef]
12. Muneta, Y.; Yoshikawa, T.; Minagawa, Y.; Shibahara, T.; Maeda, R.; Omata, Y. Salivary IgA as a useful non-invasive marker for restraint stress in pigs. *J. Vet. Med. Sci.* **2010**, *72*, 1295–1300. [CrossRef]
13. Skandakumar, S.; Stodulski, G.; Hau, J. Salivary IgA: A possible stress marker in dogs. *Anim. Welf.* **1995**, *4*, 339–350.
14. Svobodová, I.; Chaloupková, H.; Končel, R.; Bartoš, L.; Hradecká, L.; Jebavý, L. Cortisol and secretory immunoglobulin A response to stress in German shepherd dogs. *PLoS ONE* **2014**, *9*, e90820. [CrossRef]
15. Guhad, F.A.; Hau, J. Salivary IgA as a marker of social stress in rats. *Neurosci. Lett.* **1996**, *216*, 137–140. [CrossRef]
16. Kreutz, G.; Bongard, S.; Rohrmann, S.; Hodapp, V.; Grebe, D. Effects of choir singing or listening on secretory immunoglobulin A, cortisol, and emotional state. *J. Behav. Med.* **2004**, *27*, 623–635. [CrossRef]
17. McCraty, R.; Atkinson, M.; Rein, G.; Watkins, A.D. Music enhances the effect of positive emotional states on salivary IgA. *Stress Med.* **1996**, *12*, 167–175. [CrossRef]
18. Gourkow, N.; LaVoy, A.; Dean, G.A.; Phillips, C.J. Associations of behaviour with secretory immunoglobulin A and cortisol in domestic cats during their first week in an animal shelter. *Appl. Anim. Behav. Sci.* **2014**, *150*, 55–64. [CrossRef]
19. Good Scientific Practice. Available online: https://www.vetmeduni.ac.at/fileadmin/v/z/mitteilungsblatt/richtlinien/GoodScientificPractice_20140131.pdf (accessed on 10 July 2019).
20. Held, S.D.E.; Spinka, M. Animal play and animal welfare. *Anim. Behav.* **2011**, *81*, 891–899. [CrossRef]
21. Kiley-Worthington, M.; de la Plain, S. *The Behavior of Beef Suckler Cattle (Bos Taurus)*, 1st ed.; Birkhäuser Basel: Basel, Switzerland; Boston, MA, USA; Basel Springer AG: Stuttgart, Germany, 1983; p. 100.
22. Reinhardt, V. *Untersuchung zum Sozialverhalten des Rindes*; Birkhäuser Verlag: Basel, Switzerland; Boston, MA, USA; Stuttgart, Germany, 1980; pp. 40–61.
23. Benham, G.; Nash, M.R.; Baldwin, D.R. A comparison of changes in secretory immunoglobulin A following a stress-inducing and stress-reducing task. *Stress Health* **2009**, *25*, 81–90. [CrossRef]

24. Jensen, M.B.; Vestergaard, K.S.; Krohn, C.C. Play behavior in dairy calves kept in pens: The effect of social contact and space allowance. *Appl. Anim. Behav. Sci.* **1998**, *56*, 97–108. [CrossRef]
25. Martin, P.; Bateson, P. *Measuring Behaviour: An Introductory Guide*; Cambridge University Press: Cambridge, UK, 2007.
26. R Core Team. *R: A Language and Environment for Statistical Computing*; R Foundation for Statistical Computing: Vienna, Austria, 2015.
27. Hucklebridge, F.; Clow, A.; Evans, P. The relationship between salivary secretory immunoglobulin A and cortisol: Neuroendocrine response to awakening and the diurnal cycle. *Int. J. Psychophysiol.* **1998**, *31*, 69–76. [CrossRef]
28. Eriksson, E.; Royo, F.; Lyberg, K.; Carlsson, H.E.; Hau, J. Effect of metabolic cage housing on immunoglobulin A and corticosterone excretion in faeces and urine of young male rats. *Exp. Physiol.* **2004**, *89*, 427–433. [CrossRef]
29. Kugler, J.; Hess, M.; Haake, D. Secretion of salivary immunoglobulin A in relation to age, saliva flow, mood states, secretion of albumin, cortisol, and catecholamines in saliva. *J. Clin. Immunol.* **1992**, *12*, 45–49. [CrossRef]
30. Chiappin, S.; Antonelli, G.; Gatti, R.; De Palo, E.F. Saliva specimen: A new laboratory tool for diagnostic and basic investigation. *Clin. Chim. Acta* **2007**, *383*, 30–40. [CrossRef]
31. Humphrey, S.P.; Williamson, R.T. A review of saliva: Normal composition, flow, and function. *J. Prosthet. Dent.* **2001**, *85*, 162–169. [CrossRef]
32. Dillon, K.M.; Minchoff, B.; Baker, K.H. Positive emotional states and enhancement of the immune system. *Int. J. Psychiatry Med.* **1985**, *15*, 13–18. [CrossRef]
33. McKune, A.J.; Starzak, D.; Semple, S.J. Repeated bouts of eccentrically biased endurance exercise stimulate salivary IgA secretion rate. *Biol. Sport* **2015**, *32*, 21–25. [CrossRef]
34. Gillum, T.L.; Kuennen, M.; Gourley, C.; Schneider, S.; Dokladny, K.; Moseley, P. Salivary antimicrobial protein response to prolonged running. *Biol. Sport* **2013**, *30*, 3–8. [CrossRef]
35. Rutherford, K.M.D.; Donald, R.D.; Lawrence, A.B.; Wemelsfelder, F. Qualitative Behavioural Assessment of emotionality in pigs. *Appl. Anim. Behav. Sci.* **2012**, *139*, 218–224. [CrossRef]
36. Wickham, S.L.; Collins, T.; Barnes, A.L.; Miller, D.W.; Beatty, D.T.; Stockman, C.; Blache, D.; Wemelsfelder, F.; Fleming, P.A. Qualitative behavioral assessment of transport-naïve and transport-habituated sheep. *J. Anim. Sci.* **2012**, *90*, 4523–4535. [CrossRef]
37. Estep, D.Q.; Hetts, S. Interactions, relationships and bonds: The conceptual basis for scientist-animal relations. In *The Inevitable Bond—Examining Scientist-Animal Interactions*; Davis, H., Balfour, D., Eds.; Cambridge University Press: Cambridge, UK, 1992; pp. 6–26.
38. Waiblinger, S.; Boivin, X.; Pedersen, V.; Tosi, M.; Janczak, A.M.; Visser, E.K.; Jones, R.B. Assessing the human-animal relationship in farmed species: A critical review. *Appl. Anim. Behav. Sci.* **2006**, *101*, 185–242. [CrossRef]
39. Ahloy-Dallaire, J.; Espinosa, J.; Mason, G. Play and optimal welfare: Does play behavior indicate the presence of positive affective states? *Behav. Process.* **2018**, *156*, 3–15. [CrossRef]
40. Hausberger, M.; Fureix, C.; Bourjade, M.; Wessel-Robert, S.; Richard-Yris, M. On the significance of adult play: What does social play tell us about adult horse welfare? *Naturwissenschaften* **2012**, *99*, 291–302. [CrossRef]
41. Deinzer, R.; Kleineidam, C.; Stiller-Winkler, R.; Idel, H.; Bachg, D. Prolonged reduction of salivary immunoglobulin A (sIgA) after a major academic exam. *Int. J. Psychophysiol.* **2000**, *37*, 219–232. [CrossRef]
42. Lürzel, S.; Götz, J.; Stanitznig, A.; Patzl, M.; Waiblinger, S. Salivary s-IgA as an indicator of positive emotions in cattle? In Proceedings of the WAFL 2014—6th International Conference on the Assessment of Animal Welfare at Farm and Group Level, Clermont-Ferrand, France, 3–5 September 2014; p. 118.

 © 2019 by the authors. Licensee MDPI, Basel, Switzerland. This article is an open access article distributed under the terms and conditions of the Creative Commons Attribution (CC BY) license (http://creativecommons.org/licenses/by/4.0/).

Article

Intra- and Inter-Observer Reliability of Qualitative Behaviour Assessments of Housed Sheep in Norway

Sofia Diaz-Lundahl [1,*,†], Selina Hellestveit [2,†], Solveig Marie Stubsjøen [3], Clare J. Phythian [4], Randi Oppermann Moe [1] and Karianne Muri [1]

1. Department of Production Animal Clinical Sciences, Faculty of Veterinary Medicine, Norwegian University of Life Sciences, P.O. Box 369 Sentrum, 0102 Oslo, Norway
2. Department of Basic Sciences and Aquatic Medicine, Faculty of Veterinary Medicine, Norwegian University of Life Sciences, P.O. Box 369 Sentrum, 0102 Oslo, Norway
3. Department of Animal Health and Food Safety, Norwegian Veterinary Institute, P.O. Box 750 Sentrum, 0106 Oslo, Norway
4. Section for Small Ruminant Research, Department of Production Animal Clinical Sciences, Faculty of Veterinary Medicine, Norwegian University of Life Sciences, 4325 Sandnes, Norway
* Correspondence: sofia.lundahl@nmbu.no
† These authors contributed equally to study design, analysis, conduct and reporting.

Received: 27 June 2019; Accepted: 13 August 2019; Published: 17 August 2019

Simple Summary: Qualitative behaviour assessment (QBA) is a whole-animal approach to measuring animal welfare, based on observing the animal's body language and behaviour. The method is used in different animal welfare protocols such as the Welfare Quality® (WQ®) protocols developed for poultry, cattle and swine and the AWIN protocols for sheep and goats. In Norway, farmed sheep are typically housed during the winter period for approximately six months and this presents specific risks for animal welfare, as well as specific opportunities for improvement. A welfare protocol for sheep managed under Norwegian housing conditions was developed for the Norwegian Sheep House (*FåreBygg*) project. In this study, we tested the reliability of QBA as developed for this protocol, when used by six trained observers with different professional background and experience, using video recordings. Intra-observer reliability was assessed by viewing the videos twice with a one-week interval between viewings. The statistical analyses revealed high agreement between all observers, and between scorings of the same observers at different time points. The results suggest that the tested protocol is reliable for assessing video recordings of sheep behaviour when applied by trained observers, regardless of their occupation with differing experiences of sheep health, welfare and production.

Abstract: This study tested the reliability of a Qualitative Behavioural Assessment (QBA) protocol developed for the Norwegian Sheep House (*FåreBygg*) project. The aim was to verify whether QBA scores were consistent between different observers, i.e., inter-observer reliability, and between scorings of the same observers on different time points, i.e., intra-observer reliability. Six trained observers, including two veterinary students, two animal welfare inspectors and two sheep farmers observed sheep in 16 videos, and independently scored 14 pre-defined behavioural descriptors on visual analogue scales (VAS). The procedure was repeated one week after the first scoring session. QBA scores were analysed using Principal Component Analysis. Inter- and intra-observer agreement was assessed using Kendall's coefficient of concordance (W). Principal component 1 (PC 1) and 2 (PC 2) combined explained >60% of the total variation in the QBA scores in both scoring sessions. PC 1 (44.5% in sessions 1 and 2) ranged from the positive descriptors calm, content, relaxed and friendly to the negative descriptors uneasy, vigilant and fearful, and was therefore labelled mood. PC 2 (18% in session 1, 16.6% in session 2) ranged from bright to dejected and apathetic, and was therefore labelled arousal. Kendall's coefficient of concordance of PC 1 for all observers was high in the two scoring sessions (W = 0.87 and 0.85, respectively), indicating good inter-observer reliability.

For PC 2, the agreement for all observers was moderate in both video sessions (W = 0.45 and 0.65). The intra-observer agreement was very high for all observers for PC 1 (W > 0.9) except for one, where the agreement was considered to be high (W = 0.89). For PC 2, Kendall's coefficient was very high for the veterinary students and interpreted as moderate for the two farmers and welfare inspectors. This study indicates that the QBA approach and the terms included in the *Fårebygg* protocol were reliable for assessing video recordings of sheep behaviour when applied by trained observers, regardless of whether they were a veterinary student, animal welfare inspector or sheep farmer. Further work is needed to examine the reliability of the QBA protocol when tested on-farms for sheep managed under Norwegian housing systems.

Keywords: sheep; qualitative behaviour assessment (QBA); welfare assessment protocol; observer reliability; housing; animal welfare

1. Introduction

In many developed countries, consumers of meat, milk, eggs and other products from animals are paying increasing attention to animal welfare, thus creating a demand for animal welfare-friendly products [1–3]. Consequently, there is an increasing need for the assessment. Standardized animal welfare assessment protocols incorporating valid, reliable and feasible welfare indicators provide a means of comparing the welfare of animals managed across different farms in a transparent, fair and consistent manner [4].

The more recently developed protocols typically consist of a mixture of animal-based and resource-based measurements [5]. Resource-based measurements focus on what is provided to the animal, including the environment, e.g., feed quality, spacing, hygiene and relative humidity. Animal-based indicators, on the other hand, reflect the response of an animal to a specific situation or environment [6,7]. Animal-based welfare outcomes provide a better insight into how different production and management systems influence and impact animal welfare compared to measures of resource and management inputs. Traditionally, measures of good animal health have been interpreted as a reflection of good animal welfare. For example, body condition score (BCS) has been identified as a valid, reliable and feasible animal-based indicator of sheep health and welfare [8–10]. However, a sheep in an ideal body condition is not necessarily experiencing a state of emotional well-being. Hence when measured alone, BCS and other measures of the animals' physical condition do not provide sufficient information to draw conclusions about the actual welfare state of the animal. By comparison, qualitative behaviour assessment (QBA) is a whole-animal approach, based on observing the animal's behaviour and body language [11]. One of the advantages of QBA is that the expression of positive emotions are taken into consideration, as opposed to only focusing on the presence or absence of signs of disease, pain or suffering [11,12].

QBA involves the use of behavioural descriptors, such as fearful, social, uneasy and content, to describe the welfare of an individual animal or a group of animals [13]. Originally, QBA was based on free choice profiling (FCP), whereby observers generated their own terms to describe the behaviour of animals, and then scored the intensity of each term along a visual analogue scale (VAS) [14,15]. This approach requires that the data are analysed using generalised Procrustes analysis (GPA). Other users of QBA have developed a fixed-list (FL) of predefined terms for the species of interest. Observers then score each of the terms on a visual analogue scale (VAS) (e.g., Welfare Quality Protocol for cattle). To date, principal components analysis (PCA) has been the most common approach to analyse data from fixed-list QBA, as it reveals the underlying structure of the data and reduces the number of variables to a few main dimensions, which can be interpreted in terms of animal welfare [12,16–18].

In recent years, QBA has been included in several welfare protocols such as the Welfare Quality® (WQ®) protocols developed for poultry, cattle and swine [19] and in the AWIN protocols for sheep [20]

and goats [21]. Various studies have investigated the validity of QBA, based on both FCP [22,23] and the FL approach [24]. For example, Stockman and colleagues [22] found that cattle that were assessed as agitated, restless and stressed during transport also had increased heart rate, core body temperature and plasma glucose compared to before transport. Whilst in a study of sheep, Wickham and colleagues [23] found that sheep described as more active and alert compared to the control when subjected to different stressors, showed increases in core body temperature, leptin concentration and haematocrit. The described changes in physiological parameters in both studies are consistent with a stress response, and supports the validity of the method in detecting complexities of behavioural expressions of animals.

Previous studies have found varying levels of reliability when QBA was performed on cattle using FCP [24,25], on pigs using FCP [26], and FL [12]. Methods of assessment used to test observer reliability vary from video-scoring [12] to direct observations of animals on-farm [24,25] or in experimental conditions [26]

Several reliability studies have been conducted on QBA for small ruminants [16,17,27]. In 2013, the first reliability study of QBA on sheep, based on observation of video clips found high inter-observer reliability [27].

QBA is based on the ability to interpret an animal's body language, and has been suggested to be a supplementary welfare screening tool [15]. Studies on sheep [18] and goats [16], suggest that QBA is sensitive to detecting between-farm differences in animal health and welfare outcomes. To assess whether QBA is reliable and is externally valid for application by different stakeholders with different backgrounds in sheep health and welfare, it is important that the method is tested by the range of potential on-farm users, such as farmers, veterinarians and farm animal welfare inspectors. The method also needs to be tested under the range of management and farming types found in the region of interest.

In Norway, farmed sheep are typically housed for approximately six months during the cold season (October to April) and this presents opportunities for close inspection of sheep behaviour and stockperson attention. However, there may also be specific welfare risks associated with a long period of indoor housing. To date, the reliability of QBA when applied by a range of stakeholders for assessing sheep managed under Norwegian housing conditions has not been explored.

The background to this study was that QBA was one of a number of animal-based indicators that were included in an on-farm welfare assessment protocol used in the SheepHouse (FåreBygg) project. The project had a broader aim of examining the on-farm welfare of sheep managed in the different housing systems found in Norway using a combination of valid, reliable and feasible animal- resource- and management-based indicators (Norwegian Research Council Project nr. 225353/E40).

The objective of the present study was to examine the inter- and intra-observer reliability of the QBA protocol for housed sheep managed under Norwegian housing conditions, when a group of farmers, veterinary students and farm animal welfare inspectors applied the method to video recordings.

2. Materials and Methods

2.1. Experimental Design

2.1.1. Behavioural Descriptors

The behavioural descriptors used in this study were modified from a previous QBA protocol for housed sheep, that included 12 descriptors created and used by Muri and Stubsjøen [17]. A focus group including five veterinarians with practical experience in sheep behaviour and welfare assessments, and two final year veterinary students discussed the descriptors in various steps; after watching both videos of sheep and sheep on-farm. This involved the inclusion and testing of new descriptors. Proposed changes were based on consensus about the definition, judgement and feasibility, as well as the term's meaning for sheep welfare. All terms were tested on-farm followed by new discussions and revision of the list of terms. For each descriptor, the final protocol contained a written definition that the group had agreed upon.

Fourteen behavioural descriptors were finally included in this study (Table 1), of which half were negative terms, and the other half were positive terms. One behavioural descriptor, trustful, was excluded from the reliability study because there were no stockpeople present in most videos, making the assessment of trustfulness impossible. However, the term was kept in the *FåreBygg* protocol used on-farm [28]. For the remainder of this paper we only refer to the 13 descriptors included in the reliability study.

Table 1. The definitions used for the behavioural descriptors in this study.

Behavioural Descriptor	Definition from Protocol
Calm	• Not nervous • No noticeable reaction to the presence of humans or other animals • Does not apply to animals that are sick (lethargic)
Vigilant	• Tense, nervous, shy or timid • Stops ruminating
Friendly	• Sociable • Involved in positive social interactions; Social grooming, sniffing other animals, rubbing against one another, lying side by side *and/or* • More subtle social interactions, harmonic, positive interest in each other, positive synchronous behaviour • Absence of negative social interactions
Fearful	• Animals that are showing obvious signs of fear • Trying to flee, panting or withdrawing to the far corner/end of the pen
Apathetic	• Indifferent, lack of interest in surroundings and/or discouraged • Not very responsive • The term has a more negative meaning than «dejected»
Aggressive	• Actively engaged in physical conflict with other animals
Content	• Satisfied and relaxed or engaged in positive activities • For example: Playful, social grooming, eating (if there is no competition for feed or displacements) or resting
Irritable	• Frustrated, ruthless, grumpy, threatening
Curious	• Showing positive interest and anticipation (not vigilant or watchful), and appearing explorative • Animals standing still and showing interest have not stopped ruminatingDoes not include the animals that are observing you neutrally
Uneasy	• Stressed, apprehensive, restless or impatient
Bright	• Awake and alert, showing positive interest in surroundings • Animals that are lethargic or have laboured breathing due to gestation will influence this score negatively
Dejected	• Animals that are obviously ill or depressed • Lack of interest in surroundings • Not very responsive • Lethargic • In pain
Trustful	• Showing positive interest in the observer, relaxed, not trying to flee • Are not disturbed by the presence of the observer
Relaxed	• Resting, either while standing up or lying down • Ruminating with a relaxed eye expression (heavy eyelids), not paying much attention to the observer (not curious or vigilant).

2.1.2. Observers

Six observers with different backgrounds participated in the study. Two were final year veterinary students (authors S.D-L. and S.H.) of the Norwegian University of Life Sciences (NMBU), two were experienced full-time sheep farmers and two were experienced animal welfare inspectors and veterinarians from the Norwegian Food Safety Authority (NFSA). The students had received more training in using QBA than the other observers, as they had participated in group discussions and on-farm testing for development of the QBA protocol used in this study. The veterinarians from the NFSA were familiar with the method from a former study the previous year [17]. The farmers had not received any training in the use of QBA prior to the introductory presentation given at the first video scoring session.

2.1.3. Videos

All, except for four, video recordings used in this study included recordings of adult sheep in the mid-east of Norway during the winter indoor-housing period. Videos of groups of sheep were recorded on 11 farms in connection with an earlier study [10] during a two-week period in April and May 2007. Average flock size on these farms was 111 (range 35–240), and the Norwegian White Sheep breed was kept on all but one farm that also had the old Norwegian breed *Spælsau*. The videos were selected by two of the scientists (S.M.S. and K.M.), and fourteen two-minute clips were chosen from this material, aiming to cover a variety of behavioural expressions observed in Norwegian sheep housing systems. To supplement the videos from Norwegian farms, four additional videos belonging to Professor Françoise Wemelsfelder from Scotland's Rural College (SRUC) were used (videos number 2, 6, 14 and 16). The sheep on these videos were Scottish Blackface (clips 2, 6 and 16) or mule sheep (clip 16). In total, 16 videos were included in the reliability study. A brief description of each video is presented in Table 2.

Table 2. Description of the videos used in this study. The order of presentation was different in the two scoring sessions.

Video Number	Video Description	Order of Presentation in Trial	
		1st Session	2nd Session
1	A group of ewes and lambs in an indoor pen. Most of the animals are lying down and ruminating, a few are walking about.	1	5
2	Sheep walking/running in an outdoor pen. A stockperson is moving the animals using a long stick.	2	14
3	A small pen in a sheep house where the ewes are eating and the lambs are in the background.	3	2
4	A small pen in a sheep house where a stockperson is giving concentrate to the ewes in a pen, and lambs are running behind the ewes.	4	17
5	Three adult pregnant ewes in a small indoor pen.	5	4
6	A sheep in a small indoor pen.	6	16
7	The same herd and section in the sheep house as in video 1. A stockperson is distributing hay on the floor.	7	9
8	Ewes and lambs in an indoor pen. Most of them are standing up, a few are lying down.	9	12
9	Same farm and same position as in video 1 and 7. All the ewes are eating concentrate, while the lambs are running around.	10	3
10	Ewes and lambs in an indoor pen. Most are sleeping, some are resting.	11	8
11	Ewes in an indoor pen. Most are lying down and ruminating with their eyes closed.	12	10

Table 2. Cont.

Video Number	Video Description	Order of Presentation in Trial	
		1st Session	2nd Session
12	Ewes in a large indoor pen eating hay. A couple of animals are moving around behind the others, trying to get a place by the feeding trough.	13	6
13	Ewes and lambs in a small indoor pen. The ewes are either lying down or standing still, while some of the lambs are walking/jumping about.	14	13
14	One adult ewe and two lambs in a field. The ewe is lying down, and the lambs are holding their heads against hers.	15	7
15	Same farm as in videos 1, 7 and 10. The animals are either walking around, eating or interacting with each other.	16	11
16	Adult sheep walking in a shed with straw bedding. They suddenly stop walking and some of them lower their heads quickly.	17	15

2.1.4. Study Procedure

The video sessions took place at NMBU Sandnes, section for small ruminant research, in February 2015. Before scoring started, an introductory presentation was provided by the test leader and the first authors. The behavioural terms and definitions (Table 1) were presented to the group of observers and discussed for 15 min to provide clarity. The written definitions were available to the observers throughout the period of the test. Each observer received 16 scoring sheets (one for each video) with a 125 mm visual analogue scale (VAS) from minimum to maximum below each descriptor. Minimum was defined as the level where an expression is not present among the observed animals at all, whereas maximum is the level at which the expression is dominant across the entire group being observed. The observers marked along the scale for each term, answering the question "How dominant is this behavioural term among the observed animals during the observational period?"

In 15 of the 16 videos, the observers were asked to observe all of the sheep that were visible during the video clip. In one video, they were asked to focus their observations on a group of three adult sheep (clip number 16). After each video, the observers had a few minutes to score each behavioural descriptor on the separate VAS provided for each clip. The same procedure was repeated one week later in order to assess intra-observer reliability, but the order of the video clips was changed using random number allocation.

2.2. Data Management and Statistical Analysis

VAS data were transferred to a spreadsheet (Microsoft Office Excel® 2010) by recording the distance in millimetres from the minimum mark to the point where the scale was ticked.

Data were analysed in STATA SE/12.1 (StataCorp, College Station, TX, USA).

PCA based on a correlation matrix (no rotation) was conducted on the data from the two scoring sessions. The results from PCA are presented as component scores, which describe the total variation of data in the main dimensions. The first two principal components from the analyses (PC 1, PC 2) were retained for further analysis in both sessions, based on a combination of the elbow plot criterion and Kaiser's criterion [29].

Subsequently, the component scores for the two principle components were used to assess the agreement between observers (inter-observer reliability) in the two scoring sessions, and for each observer on different days (intra-observer reliability). Agreement was assessed using Kendall's coefficient of concordance (W). The results were interpreted according to Martin and Bateson [30], where $W = 0.0–0.2$ indicates a slight correlation, $W = 0.2–0.4$ a low correlation, $W = 0.4–0.7$ a moderate correlation, $W = 0.7–0.9$ a high correlation, and $W = 0.9–1.0$ indicates a very high correlation.

As different observers had different backgrounds and experience with the method, the inter-observer reliability was also investigated within the following pairs or groups: (1) veterinary students, (2) animal welfare inspectors from NFSA, (3) sheep farmers, (4) veterinary students and animal welfare inspectors from NFSA combined, and (5) all observers combined.

3. Results

Principal component 1 and 2 combined explained >60% of the total variation in the QBA scores in both scoring sessions (62.5% in session one and 61.1% in session two). In the first scoring session, PC 1 (eigenvalue 5.8) explained 44.5% of the variation while PC 2 (eigenvalue 2.3) explained 18.0% of the variation. In the second scoring session, the variation explained was 44.5% for PC 1 (eigenvalue 5.8) and 16.6% for PC 2 (eigenvalue 2.2).

The loadings of each behavioural descriptor along the PC 1 and PC 2 from the first and second scoring session are shown in Figures 1 and 2, respectively. In the loading plot for all of the observers, PC 1 ranges from the positive descriptors calm, content, relaxed and friendly to the negative descriptors uneasy, vigilant and fearful, and was therefore labelled mood. PC 2 ranged from bright to dejected and apathetic (Figure 1), and was therefore labelled arousal.

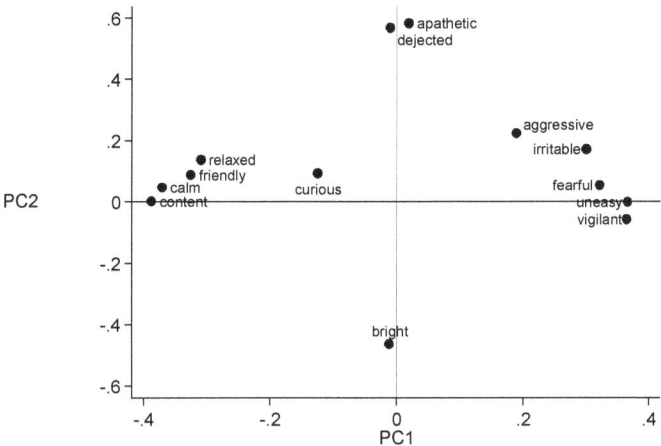

Figure 1. Loading plots generated by the first scoring session for all observers. The x-axis represents principal component 1 and the y-axis represents principal component 2.

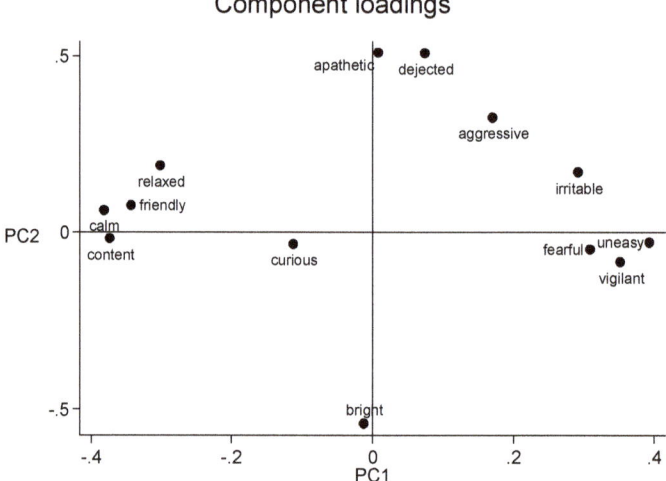

Figure 2. Loading plots generated by the second scoring session for all observers. The x-axis represents principal component 1 and the y-axis represents principal component 2.

3.1. Inter-Observer Reliability

Table 3 presents W values and p-values from the analyses of the inter-observer reliability within the different observer groups and overall. For PC 1, W values were >0.75 for all observer groups, in both video scoring sessions. For PC 2, the agreement ranged from W = 0.45 (all observers, session 1) to W = 0.91 (veterinary students, session 1).

Table 3. Inter-observer agreement given as Kendall's coefficient of concordance (W) for five groups of observers in two different scoring sessions. The groups were constructed according to background and the level of experience with QBA. PC = principal component.

Session	Group	PC 1		PC 2	
		W	p	W	p
1	Veterinary students	0.96	0.0175	0.91	0.0270
	NFSA inspectors	0.91	0.0270	0.55	0.3468
	Veterinary Students and NFSA inspector	0.86	0.0000	0.64	0.0008
	Sheep farmers	0.95	0.0191	0.82	0.0540
	All observers	0.87	0.0000	0.45	0.0004
2	Veterinary students	0.97	0.0155	0.93	0.0229
	NFSA inspectors	0.76	0.0883	0.73	0.1109
	Veterinary Students and NFSA inspector	0.82	0.0000	0.69	0.0003
	Sheep farmers	0.96	0.0170	0.81	0.0600
	All observes	0.85	0.0000	0.65	0.0000

3.2. Intra-Observer Reliability

Table 4 illustrates that for PC 1, the intra-observer agreement was very high for all observers (W > 0.9) except for one, where the agreement was high (W = 0.89). For PC 2, Kendall's coefficient was very high (>0.9) for the vet students and interpreted as moderate for the two farmers and NFSA inspectors (ranging from 0.45 to 0.67) (Table 4).

Table 4. Intra-observer agreement for individual observers given as Kendall's coefficient of concordance (W). *p*-values are considered significant at 0.05 level. PC = principal component, NFSA = Norwegian Food Safety Authority.

Observer	PC 1		PC 2	
	W	p	W	p
Veterinary student 1	0.98	0.0136	0.93	0.0223
Veterinary student 2	0.97	0.0157	0.93	0.0229
NFSA inspector 1	0.91	0.0273	0.67	0.1708
NFSA inspector 2	0.89	0.0306	0.45	0.5603
Sheep farmer 1	0.97	0.0162	0.61	0.2468
Sheep farmer 2	0.92	0.0241	0.58	0.2940

4. Discussion

This study assessed both inter- and intra-observer reliability of a fixed list QBA protocol for sheep under Norwegian housing conditions, based on video recordings. It has been suggested earlier that sheep hide signs of distress and pain and that human observers might have difficulties interpreting their behavioural expressions [10], thus underlining the need for reliable behavioural methods for inclusion in animal welfare assessment protocols.

4.1. Dimensionality of Qualitative Behavioural Assessments

PC 1 ranged from the positive descriptors calm, content, relaxed and friendly to the negative descriptors uneasy, vigilant and fearful. This summarizes the moods expressed by the sheep in the video clips and is consistent with the general tendencies in comparable studies of sheep [17,27]. The anchoring points for PC 2 are also somewhat comparable to the anchoring points identified earlier by Phythian and co-workers [27] and Muri and Stubsjøen [17], describing the arousal (bright to dejected, and apathetic). QBA studies of other species (donkeys [31], goats [16] and cattle [24] using fixed-list approach also suggests that the two main components describes mood (PC 1) and arousal (PC 2).

4.2. Inter- and Intra-Observer Reliability

All pairs of (1) veterinary students, (2) animal welfare inspectors, (3) sheep farmers, and grouping of (4) veterinary students and animal welfare inspectors combined, and (5) all observers combined obtained acceptable inter-observer reliability for PC 1. The observer pairs 1) vet students and 3) farmers reached a very high level of agreement. Comparing all observers combined, showed a high level of correlation for PC 1. Previous reliability studies have shown excellent ($W > 0.9$) and high ($W > 0.7$) inter-observer reliability for sheep [27] and [17], respectivly. There is divergence in the findings for cattle ranging from high levels of between observer agreement [24] to poor [25]. In the present study, agreement between the two welfare inspectors decreased from very high in session 1 to moderate in session 2, while correlation remained very high in the two other groups in both sessions. If one or both of the welfare inspectors did not calibrate with the pre-agreed fixed definition, this could also have caused the disagreement.

The inter-observer reliability of PC 2 was interpreted as acceptable ($W > 0.7$) for the veterinary students and sheep farmers in both sessions, and for NFSA inspectors in the second session. For the all-observers group, the inter-observer reliability of PC 2 was interpreted as not acceptable ($W < 0.7$) in both sessions, suggesting that the second component may not be sufficiently reliable using the current protocol. PC 2 was interpreted as moderate ($W = 0.69$) in a study of sheep using a previous version of the same QBA-protocol [17]. This differs slightly from the earlier work of Phythian and colleagues [27], who found higher reliability for PC 2 between all assessor groups using pre-fixed terms.

The intra-observer correlation of PC 1 was >0.7 for all observers, thus considered as acceptable reliability. However, there was variation amongst the observers. Agreement for one of the animal welfare inspectors was lower than that of the other observers. The second dimension on PC 2 showed the same tendencies as in the inter-observer study, with some groups showing a non-acceptable level of reliability.

4.3. Observer Effects: Profession, Experience and Previous Training

The current study did not identify any significant differences in the reliability of QBA when applied by different groups of observers. The groups differed in their level of training or time elapsed since training, but they also represented different professions i.e., sheep farmers, veterinary students and qualified vets working as animal welfare inspectors, with a variety of different experiences and perspectives on sheep health, welfare and behaviour.

The intra-observer reliability was equally high for all individuals independent of category ("student, farmer, welfare inspector"), suggesting that QBA could be applied by these different stakeholders to score videos of sheep behaviour with a high level of individual consistency. In this study, all observers had professions or education related to animal health or welfare. However, Duijvesteijn and colleagues [12], found that even more diverse groups (pig farmers, animal scientists and urban citizens) applying QBA to assess video clips of pigs achieved equally high level of intra-observer reliability (correlations of 0.6–0.7, using a correlation circle), regardless of their prior experience with pigs.

The inter-observer reliability for PC 1 in our study was not only high within the observer groups, but also when all-observers results were combined, suggesting that the observers, independent of their professional background with farm animal health, welfare and production, had a similar way of scoring the videos. Our results are in agreement with the sheep QBA video study of Phythian and colleagues, who also found high levels of inter-observer reliability between the groups of veterinary students/veterinarians and farm assurance assessors [27].

However, groups with other professional backgrounds may put emphasis on different aspects of animal behaviour and welfare. Duijvesteijn and colleagues [12] found significant differences in the scorings of the first dimension between farmers, researchers and urban citizens, thus indicating that there was poor between-observer agreement between these stakeholders. In that study, the varied observer groups were thought to represent different views of animal welfare: farmers seemed to have a more positive interpretation of the pigs' behaviour than the two other groups in general. The current study did not indicate the same tendencies on inter-observer reliability. However, due to logistical and time resources, only two observers were represented per professional group, which limited the statistical meaningfulness and generalisability of this result.

The level of QBA training did not seem to have an effect on the reliability of PC 1. All groups received an introductory presentation of QBA and the use of VAS scales. The veterinary students had received more training in QBA than the full-time sheep farmers, but the farmers had a much broader and longer practical experience with working with sheep managed in Norway. Others have found that experience with the species of interest and training of assessors has a considerable effect on reliability. Bokkers and colleagues [25] identified lower levels of reliability in the less experienced group performing QBA of dairy cattle, whereas Andreasen and colleagues [24] found that observers with only one day of training in the method had a very high inter-observer reliability when applying the same QBA protocol as Bokkers et al. Similarly, for sheep, Phythian and colleagues [27], found high inter-observer agreement for assessors with one day of training. Whilst the present study population was too small to separate the effect of professional background from the effect of QBA training, the results might suggest that QBA is intuitive for those experienced with sheep management as well as those less experienced observer groups that receive sufficient training in the specific species of interest.

There was more variation found in between-observer reliability for the second dimension (PC 2). Veterinary students reached a very high agreement for both intra- and inter-observer reliability, while the other observer groups provided moderate to high agreement for PC 2. Flemming and colleagues [14] found similar results for sheep, and suggested that sufficient observer calibration and training, including practice in the use of VAS scales prior to video scoring, was an important factor for achieving good intra-observer reliability.

4.4. Study Limitations

Assessment of intra-observer reliability may have been limited due to the relatively short interval (one week) between the two scoring sessions. It is possible that observers replicated the last scoring rather than assessing the videos intuitively. However, guidance on studies of diagnostic reliability [32] was followed and the order of the videos was changed in the second session to reduce this effect.

Whilst a longer interval between videos may be useful, increasing the time between the sessions would also increase the risk of observer drift, i.e., the observers altering their personal understandings of the definitions unconsciously [30].

Some scientists have considered QBA to being anthropomorphic and unscientific due to the apparent subjectivity in this approach [11]. QBA uses qualitative descriptors, but that does not necessarily mean that the method is more subjective than other methods based on observer judgement [33]. In QBA, there is a qualitative element not only in the interpretation of the results, but also when making the measurements. However, this is also true for other animal-based indicators that rely on subjective assessments such as scoring the severity of skin lesions. The expression animal welfare and our judgement of it, has its derivation in the human ability to perceive and interpret complex body language and behavioral signs [11]. Due to this, the integrative nature of QBA might be a good and even essential thing. If scientists only use quantitative measures when assessing animal welfare, they might risk overlooking important information [34], since some aspects of the concept are difficult to quantify.

Tuyttens and colleagues [35] identified that presenting observers with positive or negative information prior to QBA of cattle, pigs and laying hens resulted in significant expectation bias. From studies of sheep transported in Australia, Fleming and colleagues [14] concluded that QBA was influenced by observer bias, but the comparative ranking of animals (the pattern of interpretation), using multivariate techniques like PCA was not influenced. These results suggest that observer bias does not completely change the judgement of behaviour, and that it can be reduced by ensuring that as little additional information as possible is presented to observers prior to QBA of video recordings.

In observer ratings, expectation bias may occur for other reasons, such as a change of environmental setting. Wemelsfelder and colleagues [36] found that the observers' interpretation of pig behaviour using FCP was slightly affected by digitally altering the background environment. However, the pattern of interpretation of pig behaviour was stable. Hence, it was concluded that the contextual sensitivity of the method, is unlikely to weaken the reliability of QBA in general. Expectation bias was probably reduced in this study, because the observers were not presented to other conditions on the farm such as the condition of the barn in general and surrounding environment, or the attitude of the farmer.

The current study suggests that the QBA approach employed within the *FåreBygg* protocol is reliable for video scoring of sheep under Norwegian sheep farming conditions. Previous QBA studies of sheep have reported that higher levels of observer reliability were achieved through video recording compared to on-farm QBA. Video scoring presents controlled conditions for reliability testing but cannot fully represent an on-farm situation, and creates a challenge in comparing reliability of the two approaches. However, this issue that remains for all reliability studies based on video scoring. The next step would be to test the reliability of the same QBA protocol on-farm. This might require some alterations of the method due to several reasons. For example, in our study the videos lasted for about two minutes. In the on-farm protocol of the project to which this study belongs (*FåreBygg*), QBA was assessed for 20 min. It is possible that a longer period of observation is needed during on-farm assessments in order to capture the variety of behavioural expressions in large groups of animals, sometimes divided in different pens, rooms or buildings. This also requires that the observers changes their physical position a few times, and the presence of the observer might therefore be a disturbing factor for the animals. The videos were chosen to cover a range of different dominant aspects of sheep behaviour but it is not known whether a similar level of behavioural variation would be observed during on-farm assessments.

5. Conclusions

Overall, we identified a high level of intra-observer reliability and a very high level of inter-observer reliability for PC 1, but more varying reliability for PC 2. This study concludes that the QBA approach and the terms included in the *FåreBygg* protocol were reliable for assessing video recordings of sheep behaviour when applied by trained observers, regardless of whether they were a veterinary student, animal welfare inspector or sheep farmer and their previous experience with the methodology. Further work is needed to examine whether similar levels of assessor reliability are achieved when the same terms are applied to assess sheep welfare on-farm, as part of the Fårebygg welfare assessment protocol, and whether the method is sufficiently sensitive to detect differences in the behavioural expression between different housing types, management practices and stockperson handling.

Author Contributions: Conceptualization and methodology, S.D.-L., S.H., K.M., R.O.M., S.M.S.; Investigation, S.D.-L., S.H., C.J.P.; Formal analysis, software and data curation, K.M.; Funding acquisition, project administration and supervision, K.M., R.O.M.; Resources K.M., S.M.S., C.J.P.; Writing—original draft, writing—review and editing, validation and visualization, S.D.-L., S.H., K.M., C.J.P., S.M.S., R.O.M. All authors read and approved the final manuscript.

Funding: The project *FåreBygg* received funding from the Norwegian Research Council (project nr. 225353/E40), Småfeprogrammet for fjellregionen and Animalia.

Acknowledgments: The authors are very grateful to Professor Francoise Wemelsfelder of SRUC for generously sharing two video clips used in this study. We are grateful to the inspectors from the Norwegian Food Safety Authority (NFSA) and sheep farmers for their willingness to participate in this study.

Conflicts of Interest: The authors declare no conflict of interest.

References

1. Cornish, A.; Raubenheimer, D.; McGreevy, P. What we know about the public's level of concern for farm animal welfare in food production in developed countries. *Animals* **2016**, *6*, 74. [CrossRef] [PubMed]
2. Bennett, R.M.; Blaney, R.J. Estimating the benefits of farm animal welfare legislation using the contingent valuation method. *Agric. Econ.* **2003**, *29*, 85–98. [CrossRef]
3. Blokhuis, H.J. International cooperation in animal welfare: The Welfare Quality® project. *Acta Vet. Scand.* **2008**, *50*, S10. [CrossRef]
4. Meagher, R.K. Observer ratings: Validity and value as a tool for animal welfare research. *Appl. Anim. Behav. Sci.* **2009**, *119*, 1–14. [CrossRef]
5. Welfare Assessment of Farms Animals. Available online: http://www1.clermont.inra.fr/wq/# (accessed on 18 June 2019).
6. Main, D.; Kent, J.; Wemelsfelder, F.; Ofner, E.; Tuyttens, F. Applications for methods of on-farm welfare assessment. *Anim. Welf.* **2003**, *12*, 523–528.
7. Whay, H. The journey to animal welfare improvement. *Anim. Welf.* **2007**, *16*, 117.
8. Morgan-Davies, C.; Waterhouse, A.; Pollock, M.; Milner, J. Body condition score as an indicator of ewe survival under extensive conditions. *Anim. Welf.* **2007**, *17*, 71–77.
9. Phythian, C.J.; Michalopoulou, E.; Duncan, J.S. Assessing the Validity of Animal-Based Indicators of Sheep Health and Welfare: Do Observers Agree? *Agriculture* **2019**, *9*, 88. [CrossRef]
10. Stubsjøen, S.; Hektoen, L.; Valle, P.; Janczak, A.; Zanella, A. Assessment of sheep welfare using on-farm registrations and performance data. *Anim. Welf.* **2011**, *20*, 239–251.
11. Wemelsfelder, F. How animals communicate quality of life: The qualitative assessment of behaviour. *Anim. Welf.* **2007**, *16*, 25–31.
12. Duijvesteijn, N.; Benard, M.; Reimert, I.; Camerlink, I. Same pig, different conclusions: Stakeholders differ in qualitative behaviour assessment. *J. Agric. Environ. Ethics* **2014**, *27*, 1019–1047. [CrossRef]
13. Welfare Quality® Assessment Protocol for Cattle. Available online: http://www.welfarequalitynetwork.net/media/1088/cattle_protocol_without_veal_calves.pdf (accessed on 18 June 2019).
14. Fleming, P.; Wickham, S.; Stockman, C.; Verbeek, E.; Matthews, L.; Wemelsfelder, F. The sensitivity of QBA assessments of sheep behavioural expression to variations in visual or verbal information provided to observers. *Animal* **2015**, *9*, 878–887. [CrossRef]

15. Wemelsfelder, F.; Lawrence, A.B. Qualitative assessment of animal behaviour as an on-farm welfare-monitoring tool. *Acta Agric. Scand. Sect. A Anim. Sci.* **2001**, *51*, 21–25.
16. Grosso, L.; Battini, M.; Wemelsfelder, F.; Barbieri, S.; Minero, M.; Dalla Costa, E.; Mattiello, S. On-farm Qualitative Behaviour Assessment of dairy goats in different housing conditions. *Appl. Anim. Behav. Sci.* **2016**, *180*, 51–57. [CrossRef]
17. Muri, K.; Stubsjøen, S. Inter-observer reliability of Qualitative Behavioural Assessments (QBA) of housed sheep in Norway using fixed lists of descriptors. *Anim. Welf.* **2017**, *26*, 427–435. [CrossRef]
18. Phythian, C.J.; Michalopoulou, E.; Cripps, P.J.; Duncan, J.S.; Wemelsfelder, F. On-farm qualitative behaviour assessment in sheep: Repeated measurements across time, and association with physical indicators of flock health and welfare. *Appl. Anim. Behav. Sci.* **2016**, *175*, 23–31. [CrossRef]
19. Welfare Quality Network: Assessment Protocols. Available online: http://www.welfarequalitynetwork.net/en-us/reports/assessment-protocols/ (accessed on 19 June 2019).
20. AWIN Welfare Assessment Protocol for Sheep. Available online: https://air.unimi.it/retrieve/handle/2434/269102/384790/AWINProtocolGoats.pdf (accessed on 18 May 2019).
21. AWIN Welfare Assessment Protocol for Goats. Available online: http://uni-sz.bg/truni11/wp-content/uploads/biblioteka/file/TUNI10015667(1).pdf (accessed on 18 May 2019).
22. Stockman, C.; Collins, T.; Barnes, A.; Miller, D.; Wickham, S.; Beatty, D.; Blache, D.; Wemelsfelder, F.; Fleming, P. Qualitative behavioural assessment and quantitative physiological measurement of cattle naïve and habituated to road transport. *Anim. Prod. Sci.* **2011**, *51*, 240–249. [CrossRef]
23. Wickham, S.L.; Collins, T.; Barnes, A.L.; Miller, D.W.; Beatty, D.T.; Stockman, C.A.; Blache, D.; Wemelsfelder, F.; Fleming, P.A. Validating the use of qualitative behavioral assessment as a measure of the welfare of sheep during transport. *J. Appl. Anim. Welf. Sci.* **2015**, *18*, 269–286. [CrossRef]
24. Andreasen, S.N.; Wemelsfelder, F.; Sandøe, P.; Forkman, B. The correlation of Qualitative Behavior Assessments with Welfare Quality® protocol outcomes in on-farm welfare assessment of dairy cattle. *Appl. Anim. Behav. Sci.* **2013**, *143*, 9–17. [CrossRef]
25. Bokkers, E.; De Vries, M.; Antonissen, I.; de Boer, I. Inter-and intra-observer reliability of experienced and inexperienced observers for the Qualitative Behaviour Assessment in dairy cattle. *Anim. Welf.* **2012**, *21*, 307–318. [CrossRef]
26. Wemelsfelder, F.; Hunter, T.E.; Mendl, M.T.; Lawrence, A.B. Assessing the 'whole animal': A free choice profiling approach. *Anim. Behav.* **2001**, *62*, 209–220. [CrossRef]
27. Phythian, C.; Michalopoulou, E.; Duncan, J.; Wemelsfelder, F. Inter-observer reliability of Qualitative Behavioural Assessments of sheep. *Appl. Anim. Behav. Sci.* **2013**, *144*, 73–79. [CrossRef]
28. Muri, K.; Norwegian University of Life Sciences, Oslo, Norway. Personal communication, 2019.
29. Tabachnick, B.G.; Fidell, L.S.; Ullman, J.B. *Using Multivariate Statistics*, 6th ed.; Pearson Education Limited: Essex, UK, 2013.
30. Martin, P.; Bateson, P.P.G.; Bateson, P. *Measuring Behavior: An Introductory Guide*, 3rd ed.; Cambridge University Press: Cambridge, UK, 2007; p. 176.
31. Minero, M.; Dalla Costa, E.; Dai, F.; Murray, L.A.M.; Canali, E.; Wemelsfelder, F. Use of Qualitative Behaviour Assessment as an indicator of welfare in donkeys. *Appl. Anim. Behav. Sci.* **2016**, *174*, 147–153. [CrossRef]
32. Lucas, N.P.; Macaskill, P.; Irwig, L.; Bogduk, N. The development of a quality appraisal tool for studies of diagnostic reliability (QAREL). *J. Clin. Epidemiol.* **2010**, *63*, 854–861. [CrossRef]
33. Wemelsfelder, F.; Farish, M. Qualitative categories for the interpretation of sheep welfare: A review. *Anim. Welf.* **2004**, *13*, 261–268.
34. Fraser, D. Assessing animal welfare at the farm and group level: The interplay of science and values. *Anim. Welf.* **2003**, *12*, 433–443.

35. Tuyttens, F.; de Graaf, S.; Heerkens, J.L.; Jacobs, L.; Nalon, E.; Ott, S.; Stadig, L.; Van Laer, E.; Ampe, B. Observer bias in animal behaviour research: Can we believe what we score, if we score what we believe? *Anim. Behav.* **2014**, *90*, 273–280. [CrossRef]
36. Wemelsfelder, F.; Nevison, I.; Lawrence, A.B. The effect of perceived environmental background on qualitative assessments of pig behaviour. *Anim. Behav.* **2009**, *78*, 477–484. [CrossRef]

© 2019 by the authors. Licensee MDPI, Basel, Switzerland. This article is an open access article distributed under the terms and conditions of the Creative Commons Attribution (CC BY) license (http://creativecommons.org/licenses/by/4.0/).

Article

Sow-Piglet Nose Contacts in Free-Farrowing Pens

Katrin Portele, Katharina Scheck, Susanne Siegmann, Romana Feitsch, Kristina Maschat, Jean-Loup Rault and Irene Camerlink *

Institute of Animal Welfare Science, University of Veterinary Medicine, Vienna, Veterinaerplatz 1, 1210 Vienna, Austria
* Correspondence: irene.camerlink@vetmeduni.ac.at

Received: 30 June 2019; Accepted: 25 July 2019; Published: 31 July 2019

Simple Summary: The mother–offspring interaction is important for the young's development, yet it is rarely taken into account in farm animals for whom restricted contact and early separation from the mother are common. Nose-to-nose contact is a poorly understood form of social interaction between pigs. We investigated the occurrence and type of nose-to-nose contacts and whether and why it could differ between sows and piglets. Twenty-two sows and their 249 piglets were observed in free-farrowing pens for the first three weeks of the piglets' life. Sows and their piglets made nose contact with each other every 10 min, on average. Heavier piglets made more nose contact with the sow than lighter piglets in their first week of life. Unexperienced mothers nosed their piglets more in the second week of the piglets' life. Allowing sows and piglets to freely make nose contact may improve mother–young relationships and piglets' development, possibly benefiting animal welfare and productivity.

Abstract: Nose contact is a frequent form of social behaviour in pigs, but the motivational reasons underlying this behaviour remain unclear. We investigated the frequency, direction and type of sow–piglet nosing behaviour and its association with sow and piglet traits. Social nosing behaviour was recorded by live observations and video recordings in 22 sows and their 249 piglets in free-farrowing pens once weekly during the first three weeks after farrowing (3 times 30 min of observations per litter). Piglet-to-sow nosing occurred on average 32.8 ± 2.35 times per 30 min per litter. Heavier piglets at one week of age nosed the sow more than lighter piglets ($p = 0.01$). Piglet-to-sow nosing was unrelated to the piglet's sex or teat order. Sow-to-piglet nosing occurred on average 3.6 ± 0.53 times per 30 min, and this was unrelated to litter size. Primiparous sows nosed their piglets more in the second week after farrowing. Litters in which piglet-to-sow nosing occurred more showed less variation in the expression of this behaviour across the weeks. Social nosing between sow and piglets deserves further research to understand the positive implications of this behaviour for sow and piglet welfare.

Keywords: sow; piglet; behaviour; mother–offspring; nosing; free-farrowing; positive welfare; contact; maternal care; recognition

1. Introduction

Mother–offspring contact has crucial effects on offspring development. Studies in rodents have shown that grooming and licking of the pups by the mother have a major impact on social and neurological development and the stress-coping abilities of the offspring [1–3]. In farm animal husbandry, mother–offspring contact is often greatly restricted or even absent. Separation from the young enables an increase in the commercial output of the dam, whether this is the quantity of milk for human consumption or the number of offspring weaned per year. As a consequence, farm animals are often separated from the mother far earlier than common in nature. Moreover, the use of constraining housing systems, such as farrowing crates for sows, restrict the dam's movements and thereby her ability to initiate interactions with her offspring.

Possibly as a consequence, the mother–offspring relationship has received only little research attention in farm animals (calves [4]; sheep [5]; pigs [6]), with the main focus on the negative effects of separation from the dam [7]. This has left a knowledge gap regarding naturally occurring mother–offspring interactions and the importance of allowing the expression of maternal care in farm animals.

In wild boar, sow–piglet interactions are mainly in the form of suckling and nose contact [8]. Nose contact in pigs has been suggested to be essential in facilitating mother–offspring bonding [8–10]. Social nosing may also aid in communicating needs, e.g., nutritional needs, as suggested by the positive association between nosing and milk intake [11]. The frequency of piglet-to-sow nosing has been shown to increase during the first five days of life and then progressively decline over the first four weeks of life [10,12]. This decline coincides with the time piglets in nature would start to follow their mother outside the nest [13].

Mothers can also adjust their care to their offsprings' needs. For example, ewes are more attentive when their lambs experience pain [14]. However, in pig husbandry, the majority of sows are currently housed in farrowing crates [15], which restrict the sow's movement to standing, sitting and lying without being able to turn around or freely interact with their piglets. Sows in farrowing crates have indeed fewer interactions with their piglets than sows that can move freely [16–18]. The inability of sows to properly interact with the piglets might also be, in part, a reason for savaging of piglets by the sow [19], a serious welfare concern that is mostly seen in primiparous sows, i.e., gilts [20]. A greater understanding of the role of maternal care and mother–offspring communication is, therefore, important to animal welfare.

The aim of the current study was to examine sow–piglet nosing behaviour in free-farrowing pens and its associations to sow and piglet traits such as litter size, sow parity, and piglets' weight, teat order and sex.

2. Materials and Methods

All methods and animal use within this study were approved by the institutional ethics committee of the University of Veterinary Medicine, Vienna (protocol number 05/09/2018) in accordance with Good Scientific Practice guidelines and national legislation.

2.1. Animals and Housing

Observations took place at the pig research and teaching farm of the University of Veterinary Medicine, Vienna, Austria. Twenty-two sows (Large-White × Landrace) and their litters (249 piglets) were observed over two farrowing batches five weeks apart. Batch 1 consisted of 14 sows and 160 piglets while Batch 2 consisted of eight sows and 89 piglets. Sows were mainly in their first parity (45%) or second parity (18%), and the remaining sows were in their third (9%), fifth (14%), sixth (9%) or seventh (5%) parity. Batch 1 and 2 had five primiparous sows each. Sows were moved from their group housing into one room (same room for both batches) and housed in BeFree farrowing pens (Schauer Agrotronic GmbH, Prambachkirchen, Austria) one week before expected parturition. Sows had previously also farrowed in free-farrowing pens. The BeFree pen had a floor space of 2.22 × 2.86 m in total (6.35 m^2, with 4.2 m^2 for the sow) with plastic slatted floor and a concrete lying area. Sows had a feeder with drinker and a hay rack and a bar on one side of the wall to facilitate lying down. They had ad libitum access to water and were fed dry commercial sow feed (pellets) at 7:00, 11:30 and 15:30 and received hay daily in a rack. Sows were not crated before, during or after farrowing, except when required for short-term handling of the sow or piglets. The piglets had a creep area of 1.25 × 0.61 m (0.76 m^2), one drinker (ad libitum water) and received a commercial piglet feed (pre-starter meal) from seven days of age. Lights were on between 07:00 and 16:00 and the temperature was set at 20 °C. Average litter size was 11.3 ± 0.2 (SE) piglets (range 8–13). Cross-fostering was applied if the number of piglets exceeded the number of functional teats of the sow. In the first week of life, piglets' teeth were grinded to reduce facial injuries in the piglets, but tails were kept intact. Males were castrated at 10–14 days post-partum under general anaesthesia. The piglets were ear-tagged at approximately 19 days of age.

2.2. Data Collection

Animals were observed at the end of weeks 1, 2 and 3 of lactation. Sows farrowed at most four days apart from each other, and sows were on average (means ± SE), 8.3 ± 2.1 days, 13.4 ± 1.3 days and 21.0 ± 1.5 days in lactation. In Batch 1, 12 sows were observed by live observations and two sows from the same room were observed by video recordings due to the lack of an observer owing to health reasons at those dates. To assess the potential effect of the observation method, eight sows from a previous batch (here referred to as Batch 2) were observed from video recordings. As these sows had already been weaned, it was not possible to mark, sex or weigh the piglets. Differences between the batches are shown in Table 1.

Table 1. Overview of data collection across farrowing groups (i.e., batch).

Methodology	Batch 1	Batch 2
Live observations	12 sows	-
Video observations	2 sows	8 sows
Data recording	Piglet-to-sow behaviour Sow-to-piglet behaviour Nosing during/after suckling Piglet teat order (2 to 3 time points) Litter size Piglet body weight, weekly Piglet sex	Piglet-to-sow behaviour by litter Sow-to-piglet behaviour
Data set	160 piglets 3 time points (total 480 data points) 1534 piglet-to-sow occurrences 153 sow-to-piglet occurrences	8 litters (89 piglets) 3 time points (total 24 data points) 555 piglet-to-sow occurrences 83 sow-to-piglet occurrences
Analyses	by litter by piglet	by litter

We recorded the frequency of four piglet behaviours directed toward the sow (nose-to-nose, nosing snout, nosing ear, and nosing head, Figure 1) and one sow behaviour directed toward the piglets (sow-to-piglet nosing) (Table 2). Nosing was counted if physical contact occurred. Multiple contacts performed by the same individual <5 sec apart were counted as one occurrence. We also recorded whether the nosing behaviour occurred during a suckling bout or within one minute after a suckling bout, for Batch 1 only (Table 2). Each litter was continuously observed for three blocks of 10 min each (total 30 min) in weeks 1, 2 and 3 of lactation, resulting in a total of 90 min of behavioural observations per litter. Sampling only took place when more than 50% of the piglets in a litter were active and the sow was not feeding at the beginning of observation. Otherwise, the litter was observed at the next observation time that day when the litter was active and the sow was not feeding. This resulted in 2089 occurrences of piglet-to-sow behaviour and 236 occurrences of sow-to-piglet behaviour (Table 1).

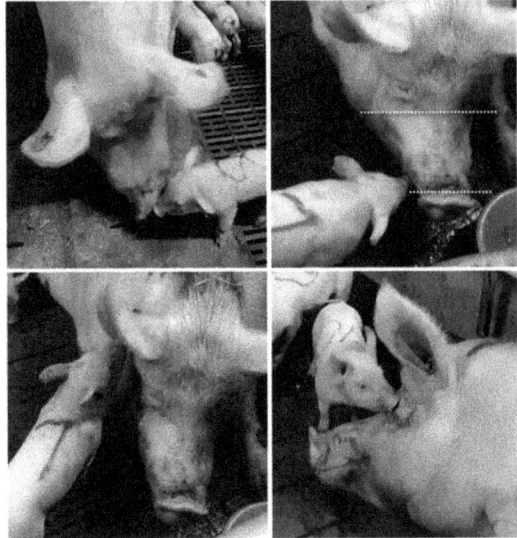

Figure 1. Target areas of the four types of piglet-to-sow nosing behaviours: nose-to-nose (top left), nosing snout, with snout area indicated between dotted lines (top right), nosing ear (bottom left), and nosing head (bottom right).

Table 2. Ethogram.

Behaviour	Description
Nose-to-nose	Piglet touches the nose disc of the sow with its nose disc
Nosing snout	Piglet touches or gently rubs the snout of the sow with its nose disc
Nosing ear	Piglet touches or gently rubs the ear of the sow with its nose disc without taking the ear into its mouth
Nosing head	Piglet touches or gently rubs the head of the sow (excluding nose disc, snout and ear) with its nose disc. Can include licking and nibbling hairs or eyelashes.
Sow-to-piglet nosing	Sow initiates contact and gently touches the piglet on any body part with her nose disc
Nosing during suckling [1]	During nursing, the piglet stops massaging the udder, or removes itself from the udder, and touches the sow's nose disc, snout, head or ear with its nose disc
Nosing after suckling [1]	Within 1 min after nursing, the piglet touches the sow's nose disc, snout, head or ear with its nose disc

[1] Behaviour only recorded for Batch 1.

Behaviours during live and video observations were recorded using the 'Ad libitum' sampling option in the app Animal Behaviour Pro version 1.2 (University of Kent, Canterbury, UK), installed on iPad or iPhone (iOS devices).

Live behavioural observations were carried out between 13:00 to 16:00 h. Piglets were marked with a marker pen for individual identification in the morning before observations, leaving at least a 30 min break before the start of the live and video observations. As piglets had no ear tag, the identification with the back number applied only to that observation day (i.e., piglets could not be re-identified across observation weeks).

Videos were recorded using overhead cameras (GV-BX 1300-KV, Geovision, Taipei, Taiwan) in a waterproof case (HEB32K1, Videotec, Schio, Italy) placed with a top-down view ~5 m above the pen. The images were recorded with 1280 × 720 pixel resolution at 30 frames per second. Recordings at the end of weeks 1, 2 and 3 of lactation, between 13:00 to 16:00 h, were selected and watched using Windows Media Player. Video observations were conducted using the same protocol as for live observations. However, piglets of Batch 2 were not marked and thus not individually identifiable. Therefore, data from Batch 2 consist of frequencies of behaviour for the litter rather than individual level. Behaviour recorded from video recordings can sometimes differ from behaviour recorded through live observations due to human presence during live observations. The sows at the research unit are selected for docile behaviour towards humans as students work with them frequently while the sows are unrestrained in a free-farrowing pen. Comparisons of the behaviour recorded through videos compared to live observations showed that there was no significant difference between the two recording methods in the frequency of sow-to-piglet contact (t-test: $t_{64} = 0.64$; $p = 0.52$) and piglet-to-sow contact (t-test: $t_{64} = 1.59$; $p = 0.12$).

The live observations were conducted by three observers simultaneously while each rotated between four litters, whereas video observations were conducted by a single observer. Inter-observer reliability was calculated for all four observers from a 10-min live observation block with the package "irr" [21] in R version 3.5.1 [22] using a two-way mixed, agreement, single-measures Intraclass Correlation Coefficient (ICC). The results showed excellent agreement for overall nosing (98.6%), as well as for the five behaviours separately (nose-to-nose: 89.6%, nosing snout: 93.4%, nosing ear: 94.1%, nosing head: 100%, sow-to-piglet nosing: 100%).

Piglets of Batch 1 ($n = 160$) were sexed and weighed in weeks 1, 2 and 3 of lactation. There were 84 males (52%) and 76 females (48%). Piglet teat order was recorded opportunistically across the observation days in Batch 1, resulting in two to three occurrences per piglet in total. As weight would be similar for piglets across the weeks, while piglet identity was unknown, only the weight at week 1 was analysed (when the piglet was the experimental unit) to avoid pseudoreplication. Average litter weight (by week) was included in the analysis of sow-to-piglet behaviour.

The data have been made available online (see Supplementary Materials statement).

2.3. Data Analysis

The coefficient of variation (CV) was calculated for the total amount of piglet-to-sow nosing of the litter and for sow-to-piglet nosing across the lactation weeks.

Statistical analyses were performed using SAS version 9.4 (Statistical Analysis Software, SAS Institute, USA). The experimental unit for the frequency of piglet-to-sow nosing and sow-to-piglet nosing was the litter.

Differences in piglet-to-sow nosing between the lactation weeks (i.e., observation weeks) were analysed at litter level using a linear mixed model (MIXED Procedure) with piglet-to-sow nosing behaviour as a response variable and, as fixed variables, the litter size and lactation week. The lactation week was included as a repeated variable with the litter nested within the batch specified as a subject to account for the repeated observations across the lactation weeks per litter.

Frequency of sow-to-piglet nosing was analysed using a linear mixed model (MIXED Procedure) with the fixed effects of sow parity (primiparous or multiparous), week of lactation (1–3), litter size, average litter weight and the interaction sow parity × week of lactation. The lactation week was included as a repeated variable with the sow nested within the batch specified as subject to account for the repeated observations across the lactation weeks per sow.

The relationships between piglet-to-sow nosing (nose-to-nose, nosing snout, nosing ear, nosing head, and total frequency) and the piglets' body weight, sex and teat order were analysed with the piglet as the experimental unit (data of Batch 1 only, $n = 160$) using linear mixed models. The fixed effects were sex (male or female), body weight week 1, average teat order, and the interaction of sex ×

body weight week 1. The litter was included as a random effect to account for the lack of independence between the piglets in the same litter.

Variables were omitted from the models if they had a *p*-value >0.10 and only if their removal improved the model fit as assessed through the Akaike information criterion (AIC) and Bayesian Information Criterion (BIC). For each model, residuals of the response variables were assessed for the normality of their distribution (Shapiro–Wilk test statistics), homogeneity and outliers. Across all variables, ten data points with frequencies greater than three standard deviations from the mean were identified as outliers. They were checked against the original data and were determined as unlikely to be recording errors and, therefore, were retained. Pearson correlations were conducted in order to assess relationships between the average frequency of piglet-to-sow nosing and sow-to-piglet nosing, and between CV and average frequency of nosing. Data are presented as Least Square means (LS-means) with standard errors (SE). *p*-values <0.05 were considered statistically significant, whereas *p*-values between 0.05 and 0.10 are stated with their actual values as tendencies.

3. Results

3.1. Piglet-To-Sow Nosing Behaviour

The frequency of piglet-to-sow nosing was, on average, 32.8 ± 2.35 occurrences per 30 min per litter, or the equivalent of 2.9 ± 0.21 times per 30 min per piglet. Piglet-to-sow nosing averaged 34.0 ± 4.05 occurrences in week one, 35.8 ± 4.05 occurrences in week two, and 27.5 ± 4.05 occurrences in week three, with no significant differences between weeks ($F_{2,42} = 1.18$; $p = 0.32$). For all weeks, the frequency with which individual piglets directed nosing to the sow was not reciprocated by the sow towards the individual piglets in the same frequency (correlation between nosing given and nosing received: $r = 0.03$; $p = 0.56$). Nosing the ear occurred least in week three ($F_{2,42} = 4.20$; $p = 0.02$), whereas nose-to-nose contact and nosing of the snout and head did not significantly differ across the lactation weeks (Figure 2). Across the three weeks, 12.45% of piglet-to-sow nosing occurred during a suckling bout, 8.15% within one minute after a suckling bout and 79.40% between suckling bouts, with no statistically significant differences between weeks.

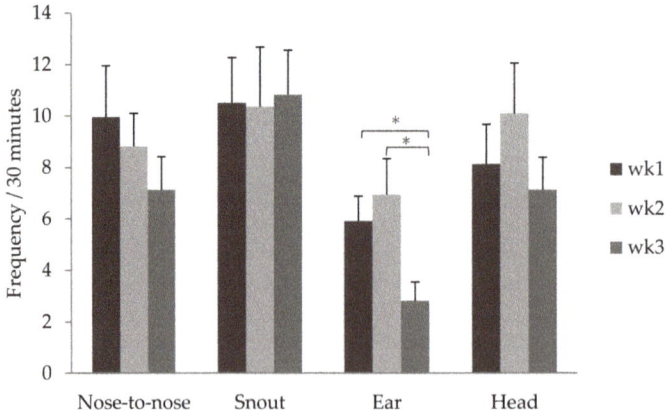

Figure 2. Average frequency of the four types of piglet-to-sow nosing behaviour ($n = 22$ litters) by observation week (30 min per week). Values are LS-means with SE. * significant difference indicated by $p < 0.05$.

In the first week of life, piglets weighed on average 3.18 ± 0.054 kg (means ± SE; range 1.32–4.68 kg), whereby heavier piglets performed more nosing toward the sow (b = 0.78 ± 0.641; $F_{1,111} = 6.18$; $p = 0.01$; Figure 3). The frequency of piglet-to-sow nosing did not differ between males (4.07 ± 0.566 occurrences) and females (3.44 ± 0.573 occurrences) ($F_{1,111} = 1.23$; $p = 0.27$) or according to the interaction of sex × body weight wk 1 ($F_{1,111} = 0.79$; $p = 0.38$). Piglet-to-sow nosing was not related to the piglet's teat order ($F_{7,111} = 0.76$; $p = 0.62$). Moreover, there was no significant relationship between piglet weight and their average teat order ($r = -0.02$; $p = 0.67$).

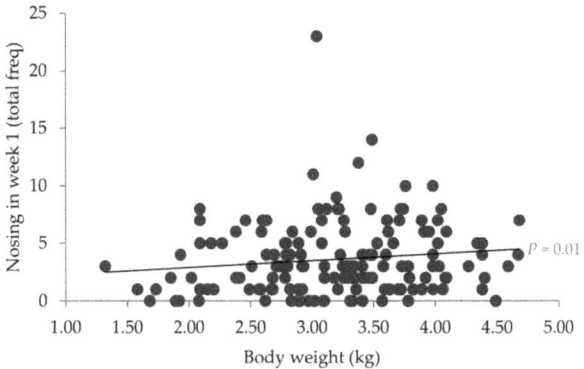

Figure 3. Relationship between piglet body weight and the frequency of piglet-to-sow nosing behaviour in the first week of life ($n = 160$ piglets).

3.2. Sow-To-Piglet Nosing Behaviour

Sows nosed their piglets on average 3.6 ± 0.53 times (range: 0–24) per 30 min. The frequency with which the sow nosed her litter did not correlate with the frequency of nosing that the piglets in the litter directed towards the sow ($r = -0.19$; $p = 0.40$). Sow-to-piglet nosing did not differ according to the number of piglets in the litter ($F_{1,11} = 0.48$; $p = 0.50$), the average litter weight ($F_{1,23} = 0.16$; $p = 0.69$) or the lactation week (week 1: 4.04 ± 0.91 occurrences, week 2: 3.30 ± 0.91 occurrences, week 3: 3.59 ± 0.91 occurrences; $F_{2,23} = 0.08$; $p = 0.92$). The frequency of sow-to-piglet nosing did not differ according to sow parity (primiparous: 4.62 ± 1.03; multiparous: 3.10 ± 0.73; $F_{1,11} = 1.26$; $p = 0.29$), but tended to differ according to the interaction of parity and lactation week ($F_{2,23} = 2.90$; $p = 0.07$; Figure 4), with primiparous sows showing significantly more sow-to-piglet nosing in the second week of lactation compared to multiparous sows ($p = 0.02$), and a trend for primiparous sows to perform more sow-to-piglet nosing in week two compared to week one ($p = 0.06$).

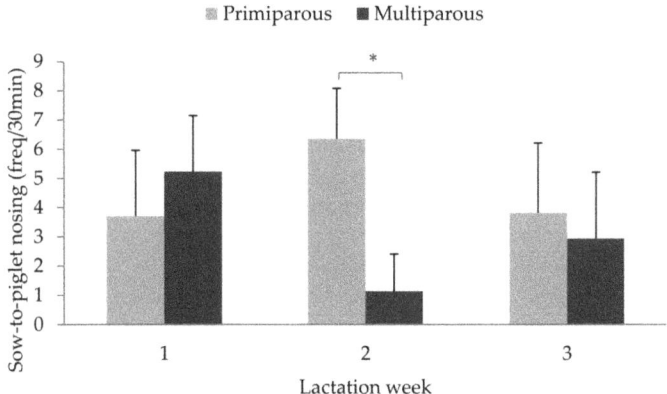

Figure 4. Frequency of sow-to-piglet nosing behaviour (per 30 min) for primiparous sows (i.e., gilts; $n = 10$) and multiparous sows ($n = 12$) throughout lactation. Values are LS-means with SE. * $p < 0.05$.

3.3. Variation in Social Nosing Behaviour Across the Lactation Period

The variation in piglet-to-sow nosing within the litter across the lactation weeks, as determined by the coefficient of variation (CV), was, on average, 40.7% ± 4.06% (Figure 5a), and ranged from 8.5% to 75.2% between litters. The CV of piglet-to-sow nosing and its average frequency were significantly negatively correlated across the lactation period ($r = -0.46$; $p = 0.03$; Figure 5a). The CV for sow-to-piglet nosing across the lactation weeks was on average 89.1% ± 10.89% (Figure 5b), and ranged from 0% to 173.2% between sows; two sows did not nose their piglets at all during observations (frequency and CV of 0; these were a primiparous and a multiparous sow from video and live observations, respectively). One primiparous sow in the live observations (most right data point) consistently nosed her piglets frequently, on average, once every 2.5 min. The CV for sow-to-piglet nosing was not correlated with its average frequency ($r = -0.13$; $p = 0.57$; Figure 5b).

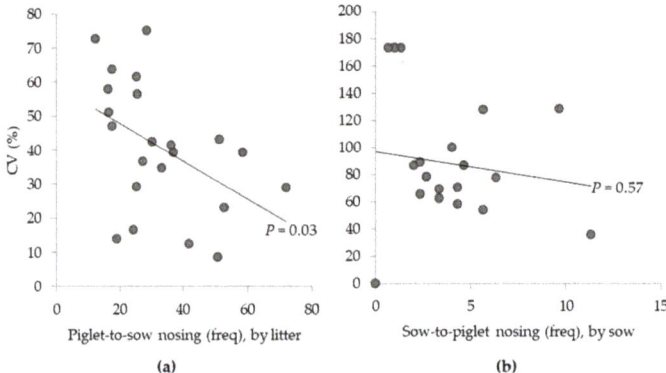

Figure 5. Coefficient of variation (CV; %) for the average frequency (per 30 min of observation) of piglet-to-sow nosing (**a**) and sow-to-piglet nosing (**b**) behaviours across the lactation period ($n = 22$ litters).

4. Discussion

Social nosing was frequently observed between the sow and her piglets, and for piglet-to-sow nosing behaviour each of the four subtypes were observed. Heavier piglets nosed the sow more than lighter piglets in the first week of life. Primiparous sows showed more nosing behaviour toward their piglets in the second week of lactation. These findings altogether suggest a role for social nosing between the sow and her piglets in social recognition, communication and maternal care.

4.1. Piglet-To-Sow Nosing Behaviour

Piglets frequently initiated nose contact with the sow. On average a piglet made a nose contact with the sow 2.9 times in the 30 min of observations. This is slightly higher than in the study of Stangel and Jensen [10], who recorded between one and two piglet-to-sow nose contacts per 2 h in piglets up to 10 days of age housed in a large outdoor enclosure (7–13 ha), and Blackshaw and Hagelsø [23] who recorded around 2.4 piglet-to-sow nose contacts per hour in free-farrowing pens (2.6 × 2.3 m). The higher occurrence of piglet-to-sow nosing in the current study may be due to different factors such as, amongst others, a larger litter size as compared to the aforementioned studies or the inclusion of nosing the ear and head rather than solely nose–nose contacts, as was the case in previous studies. Indeed, the four subtypes of piglet-to-sow nosing behaviours that we recorded were observed to be similarly prevalent, and constant over the first three weeks of life apart from piglets nosing the sow's ear, which reduced in week three. A reduction in contact was expected as piglets may divert to other activities as they get more active, and the finding that snout contacts remain stable across the weeks may be related to its wider function in communication [8].

We initially predicted that lighter-weight piglets, who may suffer from lack of milk, would initiate more social nosing with the sow to give an honest signal of their need, as they do with vocalizations [24,25]. In contrast, we found that in the first week of life, heavier piglets performed more nosing toward the sow. Heavier piglets may be healthier and stronger and have more energy left for activity, including for interacting with the sow. The causality of the relationship between body weight and nosing is, however, not necessarily unidirectional as previous work found that weaned pigs that received more nosing from others had a better growth performance [26], although in the present study, piglet-to-sow nosing did not appear to be necessarily reciprocated by the sow. Piglet-to-sow nosing has also been related to suckling and milk ejection [11,16,23]. For example, Blackshaw and Hagelsø [23] reported that 36% of the piglet-to-sow nosing was associated with lactation. Nevertheless, we did not find a difference in the frequency of nosing during or within one minute after a suckling bout in the current study.

4.2. Sow-To-Piglet Nosing Behaviour

Sows nosed their piglets on average 3.6 times per 30 min. This is again higher than reported by Stangel and Jensen [10], which may be due to the fact that in our study, nose contact toward any body part of the piglet was counted, not only nose-to-nose contact. Earlier studies in semi-natural enclosures and free-farrowing pens reported a decrease in sow-to-piglet nosing in the first ten days post-partum [10,23], and between the first and fourth week of lactation [12].

Primiparous sows delivered more nosing toward their piglets in the second week after farrowing as compared to multiparous sows, and this also tended to be higher than during their first week after farrowing. Primiparous sows can be more restless [27,28] and insecure during and after their first farrowing [29], which can increase sow-to-piglet nosing behaviour [19]. Restlessness is also associated with savaging behaviour in primiparous sows [28], which is a considerable welfare concern as it may lead to substantial piglet mortality (up to 25% mortality [19]). However, Ocepek and Andersen [30] found that sows who communicated more with their piglets, including sniffing, grunting and nudging, had a lower mortality rate in their litter, partly due to less piglet crushing. It might be that the investigation and recognition of the young [31], or other forms of maternal behaviour that sow-to-piglet nosing underlies, takes some time to establish during their first maternal experience [32]. When identification is established, the need of the sow to nose her piglets might be less. These results suggest that sow-to-piglet nosing behaviour is a form of maternal care, possibly facilitating the formation of the mother–offspring bond, which sows learn from experience with their first litter and they retain and retrieve this skill more quickly in subsequent farrowing episodes.

The amount of nosing behaviour that the piglets directed towards the sow was unrelated to the amount of nosing the sow directed to her piglets, and did not show reciprocity between individual piglets and the sow within a given week. Sow-to-piglet nosing did, however, occur infrequently and more observation hours per sow would be needed to draw conclusions upon the reciprocity of nosing behaviour between the sow and her piglets. Similarly, more data on sow-to-piglet nosing would be needed to meaningfully analyse whether the sow directs more nosing towards the weaker piglets in the litter. Based on the current frequency of sow-to-piglet nosing in relation to the litter size, at least 3 h of behavioural observations per sow would be required to assess these two above-mentioned questions.

4.3. Variation in Nosing Behaviour

Considerable within- and between-litter variation was seen in the frequency of social nosing, based on the calculated CVs and a relatively large range. Litters that displayed a greater frequency of piglet-to-sow nosing showed less variation across the weeks, suggesting that piglet-to-sow behaviour is consistent during lactation at litter level. Sow-to-piglet nosing behaviour would benefit from more observation time per sow, as this behaviour occurred infrequently, and it was not observed at all for 2 out of the 22 sows, which is a similar result to those reported from Blackshaw and Hagelsø [23]. The naturally occurring variation in sow-to-piglet behaviour, which is related to mothering abilities [30], could be used to breed good mothers for instance.

5. Conclusions

Sow–piglet nose contact is a commonly occurring behaviour when sows and piglets are able to freely interact. As pointed out in several studies, sow–piglet nose contacts may be of considerable importance to the development of piglets and to allow for the expression of maternal care by the sow. We would like to emphasize the importance of this *"most perplexing behaviour"* [16] and the need to study social nosing in pigs in its own right as a key social communication behaviour. With the increase in implementation of free-farrowing pens, we encourage further research in the direction of positive sow–piglet interactions and, subsequently, its translation into practice to enhance the potential positive implications of this behaviour for piglet and sow welfare.

Supplementary Materials: Data are available online at Mendeley: Camerlink, I. (2019), "Social nosing between sows and piglets in free farrowing pens", Mendeley Data, v1 http://dx.doi.org/10.17632/3z97hv2y4x.1

Author Contributions: Conceptualization, I.C.; methodology, I.C. and J.L.R.; formal analysis, I.C. and J.L.R.; data curation, K.P., K.S., S.S., R.F. and K.M.; writing—original draft preparation, K.P., K.S., S.S., R.F. and I.C.; writing—review and editing, I.C., K.M. and J.L.R.; visualization, I.C.; supervision, I.C.

Funding: This research received no external funding.

Acknowledgments: We are grateful to Doris Verhovsek for her guidance and practical assistance at the farm.

Conflicts of Interest: The authors declare no conflict of interest.

References

1. Caldji, C.; Tannenbaum, B.; Sharma, S.; Francis, D.; Plotsky, P.M.; Meaney, M.J. Maternal care during infancy regulates the development of neural systems mediating the expression of fearfulness in the rat. *Proc. Natl. Acad. Sci. USA* **1998**, *95*, 5335–5340. [CrossRef] [PubMed]
2. Meaney, M.J. Maternal care, gene expression, and the transmission of individual differences in stress reactivity across generations. *Annu. Rev. Neurosci.* **2001**, *24*, 1161–1192. [CrossRef] [PubMed]
3. Sakhai, S.A.; Saxton, K.; Francis, D.D. The influence of early maternal care on perceptual attentional set shifting and stress reactivity in adult rats. *Dev. Psychobiol.* **2016**, *58*, 39–51. [CrossRef] [PubMed]
4. Edwards, S.A.; Broom, D.M. Behavioural interactions of dairy cows with their newborn calves and the effects of parity. *Anim. Behav.* **1982**, *30*, 525–535. [CrossRef]
5. Searby, A.; Jouventin, P. Mother-lamb acoustic recognition in sheep: A frequency coding. *Proc. R. Soc. Lond. B Biol. Sci.* **2003**, *270*, 1765–1771. [CrossRef] [PubMed]
6. Illmann, G.; Schrader, L.; Spinka, M.; Sustr, P. Acoustical mother-offspring recognition in pigs (*Sus scrofa domestica*). *Behaviour* **2002**, *139*, 487–506. [CrossRef]
7. Weary, D.M.; Jasper, J.; Hötzel, M.J. Understanding weaning distress. *Appl. Anim. Behav. Sci.* **2008**, *110*, 24–41. [CrossRef]
8. Gundlach, H. Brutfürsorge, Brutpflege, Verhaltensontogenese und Tagesperiodik beim Europäischen Wildschwein (*Sus scrofa L.*). *Ethology* **1968**, *25*, 955–995.
9. Petersen, V.; Recén, B.; Vestergaard, K. Behaviour of sows and piglets during farrowing under free-range conditions. *Appl. Anim. Behav. Sci.* **1990**, *26*, 169–179. [CrossRef]
10. Stangel, G.; Jensen, P. Behavior of semi-naturally kept sows and piglets (except suckling) during 10 days postpartum. *Appl. Anim. Behav. Sci.* **1991**, *31*, 211–227. [CrossRef]
11. Jensen, P.; Stangel, G.; Algers, B. Nursing and suckling behavior of semi-naturally kept pigs during first 10 days postpartum. *Appl. Anim. Behav. Sci.* **1991**, *31*, 195–209. [CrossRef]
12. Jensen, P. Maternal behaviour and mother—Young interactions during lactation in free-ranging domestic pigs. *Appl. Anim. Behav. Sci.* **1988**, *20*, 297–308. [CrossRef]
13. Jensen, P.; Redbo, I. Behaviour during nest leaving in free-ranging domestic pigs. *Appl. Anim. Behav. Sci.* **1987**, *18*, 355–362. [CrossRef]

14. Hild, S.; Clark, C.C.; Dwyer, C.M.; Murrell, J.C.; Mendl, M.; Zanella, A.J. Ewes are more attentive to their offspring experiencing pain but not stress. *Appl. Anim. Behav. Sci.* **2011**, *132*, 114–120. [CrossRef]
15. Barnett, J.L.; Hemsworth, P.H.; Cronin, G.M.; Jongman, E.C.; Hutson, G.D. A review of the welfare issues for sows and piglets in relation to housing. *Aust. J. Agric. Res.* **2001**, *52*, 1–28. [CrossRef]
16. Whatson, T.S.; Bertram, J.M. Some observations on mother-infant interactions in the pig (*Sus scrofa*). *Appl. Anim. Ethol.* **1983**, *9*, 253–261. [CrossRef]
17. Chidgey, K.L.; Morel, P.C.; Stafford, K.J.; Barugh, I.W. Observations of sows and piglets housed in farrowing pens with temporary crating or farrowing crates on a commercial farm. *Appl. Anim. Behav. Sci.* **2016**, *176*, 12–18. [CrossRef]
18. Singh, C.; Verdon, M.; Cronin, G.M.; Hemsworth, P.H. The behaviour and welfare of sows and piglets in farrowing crates or lactation pens. *Animal* **2017**, *11*, 1210–1221. [CrossRef] [PubMed]
19. Ahlström, S.; Jarvis, S.; Lawrence, A.B. Savaging gilts are more restless and more responsive to piglets during the expulsive phase of parturition. *Appl. Anim. Behav. Sci.* **2002**, *76*, 83–91. [CrossRef]
20. Harris, M.J.; Li, Y.Z.; Gonyou, H.W. Savaging behaviour in gilts and sows. *Can. J. Anim. Sci.* **2003**, *83*, 819–821. [CrossRef]
21. Gamer, M.; Lemon, J.; Fellows, I.; Singh, P. Irr: Various Coefficients of Interrater Reliability and Agreement; R package version 0.84. 2012. Available online: https://cran.r-project.org/web/packages/irr/irr.pdf (accessed on 26 July 2019).
22. R Core Team. *R: A Language and Environment for Statistical Computing*; R Foundation for Statistical Computing: Vienna, Austria, 2018.
23. Blackshaw, J.K.; Hagelsø, A.M. Getting-up and lying-down behaviours of loose-housed sows and social contacts between sows and piglets during day 1 and day 8 after parturition. *Appl. Anim. Behav. Sci.* **1990**, *25*, 61–70. [CrossRef]
24. Weary, D.M.; Fraser, D. Calling by domestic piglets: Reliable signals of need? *Anim. Behav.* **1995**, *50*, 1047–1055. [CrossRef]
25. Drake, A.; Fraser, D.; Weary, D.M. Parent-offspring resource allocation in domestic pigs. *Behav. Ecol. Sociobiol.* **2008**, *62*, 309–319. [CrossRef]
26. Camerlink, I.; Bijma, P.; Kemp, B.; Bolhuis, J.E. Relationship between growth rate and oral manipulation, social nosing, and aggression in finishing pigs. *Appl. Anim. Behav. Sci.* **2012**, *142*, 11–17. [CrossRef]
27. Chen, C.; Gilbert, C.L.; Yang, G.; Guo, Y.; Segonds-Pichon, A.; Ma, J.; Evans, G.; Brenig, B.; Sargent, C.; Affara, N.; et al. Maternal infanticide in sows: Incidence and behavioural comparisons between savaging and non-savaging sows at parturition. *Appl. Anim. Behav. Sci.* **2008**, *109*, 238–248. [CrossRef]
28. Wischner, D.; Kemper, N.; Stamer, E.; Hellbruegge, B.; Presuhn, U.; Krieter, J. Characterisation of sows' postures and posture changes with regard to crushing piglets. *Appl. Anim. Behav. Sci.* **2009**, *119*, 49–55. [CrossRef]
29. Gäde, S.; Bennewitz, J.; Kirchner, K.; Looft, H.; Knap, P.; Thaller, G.; Kalm, E. Genetic parameters for maternal behavior traits in sows. *Livest. Sci.* **2008**, *114*, 31–41. [CrossRef]
30. Ocepek, M.; Andersen, I.L. Sow communication with piglets while being active is a good predictor of maternal skills, piglet survival and litter quality in three different breeds of domestic pigs (*Sus scrofa domesticus*). *PLoS ONE* **2018**, *13*, e0206128. [CrossRef]
31. Horrell, I.; Hodgson, J. The bases of sow-piglet identification. 1. The identification by sows of their own piglets and the presence of intruders. *Appl. Anim. Behav. Sci.* **1992**, *33*, 319–327. [CrossRef]
32. Nowak, R.; Porter, R.H.; Lévy, F.; Orgeur, P.; Schaal, B. Role of mother-young interactions in the survival of offspring in domestic mammals. *Rev. Reprod.* **2000**, *5*, 153–163. [CrossRef] [PubMed]

© 2019 by the authors. Licensee MDPI, Basel, Switzerland. This article is an open access article distributed under the terms and conditions of the Creative Commons Attribution (CC BY) license (http://creativecommons.org/licenses/by/4.0/).

Article

Only When It Feels Good: Specific Cat Vocalizations Other Than Meowing

Jaciana Luzia Fermo [1], Maria Alice Schnaider [1], Adelaide Hercília Pescatori Silva [2] and Carla Forte Maiolino Molento [1,*]

[1] Animal Welfare Laboratory, Federal University of Paraná, Curitiba 80035-050, Paraná, Brazil; jacianafermo@gmail.com (J.L.F.); maschnaider@yahoo.com.br (M.A.S.)
[2] Department of Literature and Linguistics, Federal University of Paraná, Curitiba 80035-050, Paraná, Brazil; adelaidehpsilva@gmail.com
* Correspondence: carlamolento@ufpr.br

Received: 31 August 2019; Accepted: 14 October 2019; Published: 29 October 2019

Simple Summary: Among carnivore animals, domestic cats are those with the most extensive vocal repertoire. This is due to their social organization, nocturnal activity and long period of contact between the mother and the offspring. In order to identify vocalizations other than meowing in two different situations, a study was performed with 74 cats divided into two groups, one associated with a pleasant situation and another with an aversive situation. Only the group exposed to the positive stimulus of being offered a favorite snack produced specific vocalizations other than meowing: recognition or trill, squeak, purring and chatter. During the aversive situation of car transport, no vocalization other than meowing was observed. The present study indicates the relevance of applying the study of vocalizations to determine the state of emotional valence in cats.

Abstract: Our objective was to identify and characterize the types of vocalization other than meowing (VOM) in two contexts, a pleasant and an aversive situation, and to study the effect of the sex of the animal. A total of 74 cats (32 tom cats and 42 queens) living in the city of Curitiba, Brazil, participated in the study; in total, 68 (29 tom cats and 39 queens) were divided into two groups according to the stimulus they were exposed to: either a pleasant situation (PS), when they were offered a snack, or an aversive situation (AS), with the simulation of a car transport event. The other six animals (three tom cats and three queens) participated in both situations. Only the PS group presented VOM; of the 40 PS animals, 14 presented VOM, mostly acknowledgment or trill and squeak. No correlation was observed between vocalization and cat sex ($p = 0.08$; Pearson's Chi-Square). Results show that VOM is exclusively associated with positive situations, suggesting that these vocalizations may be relevant for understanding the valence of cat emotional state. Further studies are warranted to advance knowledge on other VOMs and on the generalization of our findings to other situations.

Keywords: cat behavior; *Felis catus*; phonetics; welfare

1. Introduction

Communication plays a central role in various aspects of animal life, such as during breeding interactions, territorial defense, parental care and anti-predator behavior in many species. Vocalization is the active generation of sounds with the use of specific organs, which propagate through acoustic signals that transmit a wide range of information about the communicator, including his or her emotional, motivational and physiological states [1,2]. A number of studies used vocalization as an analytical tool to assess animal welfare [3–9], using non-invasive monitoring systems. The vocalization events captured may then be studied with the aid of programs that perform a detailed acoustic analysis of sound waves, employing parameters such as frequency and amplitude of the signals [2], in order to classify the different types of vocalization.

A better understanding of how animals communicate their feelings is fundamental for all aspects of animal welfare, which is an ever increasing societal demand. As an essential mode for communicating feelings and emotions in the human species, we are particularly prone to understanding oral expressions, provided they occur in a manner that we have decodified. In the context of vocalizations as indicators of animal welfare, cats (*Felis catus*) are of special interest. They have lived with humans for over 10,000 years and have become one of the most popular pet species in the world, with over 600 million individuals [10,11]. Their repertoire of vocalization is more extensive as compared to other members of the Order Carnivorae [12,13], which may be explained by their social organization, nocturnal activity and long period of contact between the mother and her offspring [12]. There are studies that report up to 17 distinct cat vocal signals and there is ongoing research to understand the meaning of these sounds especially when they are aimed at humans [14].

Vocal responses may prove to be useful tools for motor, perceptual, motivational and social development investigations in the cat [15]. Cats can express affiliative feelings during human–animal or mother–kitten interactions, as well as fear and aggression in hostile situations through vocalization and body posture [16]. Cat vocalizations are generally divided into three main categories [14,17,18]: murmur patterns, vowel patterns and forced intensity patterns. The murmur vocalization corresponds to the emission of sounds produced with the mouth closed. Examples of these sounds are purring, acknowledgment or trill, calling and grunting [19]. The vowel patterns are formed with the initially open mouth that closes gradually [19]; examples of this pattern are meowing, which occurs in a variety of situations during friendly cat–human interactions [12], more often than in cat–cat interactions [13]. The squeak, which is a loud nasal hoarse sound [20,21], is also classified as a vowel pattern of cat vocalization. Patterns of forced intensity are sounds that express aggressive emotional states and are produced with an open mouth in a relatively constant position [19,20]. Grunting, howling, growling, hissing and spitting are sounds related to aggression of various kinds [22]. Another type of forced intensity sound is the chatter, which is believed to be emitted when cats visualize prey in an attempt to mimic their sound [21]. Even though there is clear importance of vocalization in cats, there are few phonetic studies on this species and these report results of critically limited numbers of individuals, types of vocalization and methods [21].

The objective of this study was to identify and characterize the types of vocalization other than meowing (VOM) in two different contexts, a pleasant and an aversive situation, and to study the effect of the sex of the animal, in order to improve human understanding of cat communication, especially regarding emotional valence. Our hypotheses are that specific types of VOM are coherently and consistently related to either negative or positive emotional valence, since the prevalence of VOM in cats suggests they present relevant functions to the animals, and that female cats are more frequent VOM emitters than male cats, due to the important role mothers play in raising kittens and the many vocalizations involved in mother–kitten interactions.

2. Material and Methods

2.1. Animals

The project was conducted with the collaboration of cat guardians, residents in the city of Curitiba, Paraná, Southern Brazil, from 30 November 2017 to 22 March 2018. We first tested 223 mixed-bred domestic cats, exposing each cat to both an aversive (AS) and a pleasant (PS) situation. Cats were then selected using the criterion of a vocalization rate equal to or greater than five events during the recording time in each situation. This criterion resulted in the selection of 74 cats (29 males and 45 females). Among them, 68 (26 males and 42 females) reached the selection criterion only in one situation, either the AS or the PS. The other six animals (three males and three females) presented the required vocalization rate when exposed to both scenarios. This allowed for the inclusion of 40 cats per group; cats in the AS group were from four months to 14 years old, and 24 were spayed females and 16 were neutered males; the 40 cats in the PS group were between one and 14 years old, and 24 were spayed females and 16 were neutered males.

Individual recordings were made with a digital camera (Sony Cyber Shot DSC-W610 14.1 megapixels, Sony do Brasil Ltda, Rua Werner Von Siemens, 111, Lapa, CEP: 05069-010, São Paulo, SP), following the instructions proposed by Yeon et al. [16], adapted to the use of a single camera. AS was produced by placing the cat in a carrying case, which was placed in the back seat of a car, where the cat was taken for a short trip in the absence of its guardian. The recording was initiated after the car started moving, but the evaluation started after the first vocal signal of the meow. The camera was attached to the carrying case with dimensions of 55 cm in height, 52 cm in width and 71 cm in length. The PS was produced by guardians offering each cat a portion of approximately 85 g of their favorite snack in their home environment. The snack was delivered to the cat only after the vocalization of the animal. In this case, the camera was placed on a tripod, facing the animal, at a distance of approximately 1.5 m from the guardian and cat pair, set and turned on prior to the start of the test session to allow the animal to adapt to the presence of the camera. Recording began shortly before the snack was offered and the evaluation was performed from the first meow vocal signal.

2.2. Vocal Measures

The audio were analyzed continuously for a period of 3 min starting from the first vocal sound emitted by the animal—for this, we considered any vocal sound, including meowing. To perform detailed vocalization analyses, audio were separated from videos using the Audacity software (version 2.1.3). The vocalization acoustic signals, which were captured at a sampling rate of 44.1 kHz and quantized at 16 bits, were stored in.wav format files on a computer Dell Inspiron 5458, with which the acoustic analyses were performed. Audio were carefully analyzed and categorized by one author—aided by interactions with other two authors for the first analyses, and the website Meowsic [23]. For the confirmation of purr detection, two authors were directly involved, since this type of vocalization is out of Praat software frequency range of detection and was very subtle in the audio recordings. Additionally, acoustic analysis of VOM events was performed with the Praat software, version 5.3.55, developed by Paul Boersma and David Weenink, at the Institute of Phonetic Sciences at the University of Amsterdam. Other authors such as Yeon et al. [16] have previously used Praat to analyze animal vocalizations.

The vocalization variables were calculated separately for each individual. Measurements taken were identification, duration, fundamental frequency (f0) and intensity of VOM.

2.3. Statistical Analysis

The comparison between treatment groups, AS and PS, was obtained through descriptive statistics, since VOM values for AS were zero. To verify the relationship between the specific VOM emissions and the sex of the cat, Pearson's Chi-Square statistical test in Microsoft Excel was used, considering a significance level of $p < 0.05$.

2.4. Ethical Note

This experiment was approved by the Ethics Committee on Animal Use (Comissão de Ética no Uso de Animais - CEUA) of the Sector of Agrarian Sciences of the Federal University of Paraná—Brazil, during the session of 2 June 2017, registered under protocol number 055/2017.

3. Results and Discussion

Vocalizations other than meowing were completely absent when cats were exposed to AS, which was a major finding of this work: when facing the specific AS studied, the only type of vocalization cats emitted was meowing. Thus, the types of VOM emitted by cats when exposed to the specific PS studied seem to be exclusively related to feelings of positive valence. Of the 40 animals (24 females and

16 males) exposed to the PS, 14 animals presented VOM (35.0%), which were recognition or trilling, purr, squeak and chatter types of vocalization. Of the 14 animals, 11 were female and three were male cats; no correlation was observed between vocalization and cat sex ($p = 0.08$). The literature provides reasons for hypothesizing a difference between sex for cat vocalization, since females play an important role in raising kittens, an activity highly related to vocal communication [12]. Thus, considering available knowledge and the statistical result observed for sex comparison ($p = 0.08$), further studies involving more animals in PS seem warranted to better understand the relationship between cat vocalization and sex. The group of cats that did present VOM did not include kittens or juveniles.

The duration, fundamental frequency and intensity of the different types of vocalization identified in this study, as well as the number and sex of cats who emitted them, are shown in Table 1.

Table 1. Average and standard deviation for duration, fundamental frequency (f0) and intensity of the different types of vocalizations other than meowing (VOM) observed during the pleasant situation (PS) by sex of the animal, according to the analyses of 74 cats facing positive and aversive situations; only cats in the positive situations expressed VOM.

Types of Vocalization	Number of Cats	Sex	Duration (Seconds)	Fundamental Frequency (Hertz)	Intensity (Decibel)
Trill	1	Male	0.34	-	52.0
	6	Female	0.32 ± 0.19	454.92 ± 89.44	56.58 ± 5.46
Squeak	1	Male	0.81	509.38	61.06
	4	Female	0.36 ± 0.28	440.47 ± 80.29	56.24 ± 7.70
Purr	2	Male	2.51 ± 2.12	-	45.78 ± 1.47
	1	Female	1.10	-	47.77
Chatter	0	Male	-	-	-
	1	Female	1.05	-	50.49

The acknowledgment/trill was emitted by six females and had an average duration of 0.32 s, an f0 of 454.92 Hz and intensity of 56.58 dB. The male feline emitted this sound, which lasted 0.34 s, had no f0 captured by the Praat program and an intensity of 52.0 dB. In the study by Schötz [21], the trill was the most common sound after the meow, presented with a duration of 0.51 if f0 of 533 Hz, and the wave image also presented the same characteristics. The trill has already been reported in the context of the friendly approach of cats to familiar people, an expression of strong bonding to their guardians, meaning a form of recognition [12,19].

The squeak sound was emitted by four females with an average duration of 0.36 s, f0 of 440.47 Hz and intensity of 56.24 dB. A male feline produced this sound, with a duration of 0.81 s, the f0 of 509.38 Hz and intensity of 61.06 dB. There are no publications of the waveform of this sound and, because it is part of the category of vocal patterns, it has been compared to the sound of meowing. No f0 was detected for this vocalization, possibly due to the distance between the camera and animal. Schötz [21] described that the meowing f0 ranges from 221 to 1185 Hz, with a mean duration of 0.42 s. According to Schötz [14], a squeak is a raspy, nasal, high-pitched and often short meow-like call, sometimes not ending with a closing mouth and indicates a friendly request, coinciding with the context of our PS.

The purr was emitted by two males with an average duration of 2.51 s and intensity of 45.78 dB. This sound was also emitted by a female with a duration of 1.10 s and an intensity of 47.77 dB. The purr f0 was not detected by the Praat program, probably because it is likely to be below 50 Hz and may be also due to the presence of background noise. According to Bradshaw et al. [13], the purr can last 2–700 s with f0 of 25–30 Hz, there is no intensity information, and it is usually associated with contact situations. Humans often interpret purring as if the cat were happy, but some cats also purr at feeding time, requesting their food [24]. The acoustic qualities of this vocalization are difficult to ignore and are a sign of care and attention solicitude, which is usually reinforced by the guardians [13]; Thus, it seems coherent that cats in the PS emitted those sounds. However, purring requires further research, especially designed for this vocalization. It is extremely subtle and our experimental design and equipment composed a setting where the possibility of undetected purring may not be excluded.

The chatter was produced by a cat in this study with a duration of 1.05 s and an intensity of 50.49 dB; no f0 was detected for this vocalization, possibly due to the distance between the camera and the animal. Schötz [21] reported a chatter average duration of 0.74 s, f0 of 400–600 Hz, with no description of its intensity. This type of vocalization may be considered an imitation of the sounds emitted by prey in order to deceive them, as for example when cats observe birds through a window [13]. In this study, a cat produced this vocalization trying to reach the snack, which for the animal may mean the search for food, with putative hunting components.

4. Conclusions

For the first time, specific vocalizations other than meowing were identified as exclusively emitted in a situation of positive emotional valence: acknowledgement or thrill, squeaking, purring and chatter, emitted by cats when exposed to a pleasant situation of receiving their favorite snack. No difference was reported between sex; however, this issue warrants further studies with a higher number of individuals. As no vocalization other than meowing was observed during the aversive situation of car transport, the vocalization types studied seem highly relevant for understanding the valence of cat emotional state. Overall, this research indicates the relevance of applying the study of sounds other than meowing for the comprehension of the responses of cats when facing different situations and emotional states, additionally advancing knowledge regarding their characterization. Further studies are warranted to advance knowledge regarding the generalization of our findings to other pleasant and aversive situations.

Author Contributions: J.L.F. Video analysis and writing of original draft. M.A.S. Conceptualization, funding acquisition, project administration, experimental design, field work, review of first draft. A.H.P.S. Experimental design, review and interpretation of acoustic parameters. C.F.M.M. Conceptualization, funding acquisition, project administration, experimental design, review and editing, supervision.

Funding: This research was supported by a doctorate scholarship from the Coordination for the Improvement of Higher Education Personnel (CAPES).

Acknowledgments: Thanks to the cats that participated in this study, through the kind and voluntary contribution of their guardians, whom we also acknowledge. We wish to acknowledge the suggestions from three anonymous reviewers, which allowed for significant improvement of this manuscript.

Conflicts of Interest: The authors declare no conflict of interest.

References

1. Grandin, T. The feasibility of using vocalization scoring as an indicator of poor welfare during slaughter. *Appl. Anim. Behav. Sci.* **1998**, *56*, 121–128. [CrossRef]
2. Friel, M.; Kunc, H.P.; Griffin, K.; Asher, L.; Collins, L.M. Acoustic signalling reflects personality in a social mammal. *R. Soc. Open Sci.* **2016**, *3*, 160178. [CrossRef] [PubMed]
3. Weary, D.; Fraser, D. Vocal response of piglet to weaning: Effect of piglet age. *Appl. Anim. Behav. Sci.* **1997**, *54*, 153–160. [CrossRef]
4. Burgdof, J.; Panksepp, J.; Moskal, J.R. Frequency-modulated 50 kHz ultrasonic vocalizations: A tool for uncovering the molecular substrates of positive affect. *Neurosci. Biobehav. Rev.* **2011**, *35*, 1831–1836. [CrossRef] [PubMed]
5. Matthews, S.G.; Miller, A.L.; Clapp, J.; Plötz, T.; Kyriazakis, I. Early detection of health and welfare compromises through automated detection of behavioural changes in pigs. *Vet. J.* **2016**, *217*, 43–51. [CrossRef] [PubMed]
6. Cordeiro, A.F.D.S.; Nääs, I.D.A.; Baracho, M.D.S.; Jacob, F.G.; Moura, D.J.D. The use of vocalization signals to estimate the level of pain in piglets. *Eng. Agric.* **2018**, *38*, 486–490. Available online: http://submission.scielo.br/index.php/eagri/article/view/180123 (accessed on 7 April 2019). [CrossRef]

7. Urrutia, A.; Martínez-Byer, S.; Szenczi, P.; Hudson, R.; Bánszegi, O. Stable individual differences in vocalisation and motor activity during acute stress in the domestic cat. *Behav. Process.* **2019**, *165*, 58–65. Available online: https://europepmc.org/abstract/med/31132445 (accessed on 19 October 2019). [CrossRef] [PubMed]
8. Nicastro, N.; Owren, M.J. Classification of domestic cat (*Felis catus*) vocalizations by naive and experienced human listeners. *J. Comp. Psychol.* **2003**, *117*, 44. Available online: https://www.ncbi.nlm.nih.gov/pubmed/12735363. (accessed on 19 October 2019). [CrossRef] [PubMed]
9. Nicastro, N. Acoustic correlates of human responses to domestic cat (*Felis catus*) vocalizations. *J. Acoust. Soc. Am.* **2002**, *111*, 2393. [CrossRef]
10. Turner, D.C.; Bateson, P. *The Domestic Cat: The Biology of Its Behaviour*, 2nd ed.; Cambridge University Press: Cambridge, UK, 2000; p. 12.
11. Driscoll, C.A.; Juliet, C.B.; Andrew, C.K.; Stephen, J.O.B. The taming of the cat. *Sci. Am.* **2009**, *300*, 68–75. Available online: https://www.scientificamerican.com/article/the-taming-of-the-cat/?redirect=1 (accessed on 19 April 2019). [CrossRef] [PubMed]
12. Bradshaw, J.W.S. *The Behaviour of the Domestic Cat*; Wallingford, Oxfordshire: Boston, UK, 1992; pp. 93–95.
13. Bradshaw, J.W.S.; Beaumont, C. The signaling repertoire of the domestic cat and its undomesticated relatives. In *the Domestic Cat—The Biology of Its Behaviour*; Turner, D.C., Bateson, P., Eds.; Cambridge University Press: Cambridge, UK, 2000; pp. 67–93.
14. Schötz, S.; Van de Weijer, J.; Eklund, R. Phonetic Characteristics of Domestic Cat Vocalisations. In *the 1st International Workshop on Vocal Interactivity in-and-between Humans, Animals and Robots*; University of Skövde: Skövde, Sweden, 2017; pp. 5–6. Available online: https://www.researchgate.net/publication/319814201_Phonetic_Characteristics_of_Domestic_Cat_Vocalisations (accessed on 14 April 2019).
15. Brown, K.A.; Buchwald, J.S.; Johnson, J.R.; Mikolich, D.J. Vocalization in the cat and kitten. *Dev. Psychobiol.* **1978**, *11*, 559–570. [CrossRef] [PubMed]
16. Yeon, S.C.; Kim, Y.K.; Park, S.J.; Lee, S.S.; Suh, Y.L.; Houpt, K.A.; Chang, H.H.; Lee, H.C.; Yang, B.G.; Lee, H.J. Differences between vocalization Evoked by social stimuli in feral cats and house cats. *Behav. Process.* **2011**, *87*, 183–189. [CrossRef] [PubMed]
17. Moelk, M. Vocalizing in the House-Cat: A Phonetic and Functional Study. *Am. J. Psychol.* **1944**, *57*, 184–205. [CrossRef]
18. Crowell-Davis, S.L.; Curtis, T.M.; Knowles, R.J. Social organization in the cat: A modern understanding. *J. Feline Med. Surg.* **2004**, *6*, 19–28. [CrossRef] [PubMed]
19. Beaver, B.V. *Comportamento Felino—Um Guia Para Veterinários*; 2 edição Editora Roca: São Paulo, Brazil, 2005; p. 372.
20. Little, S.E. *O Gato: Medicina Interna*, 1 ed.; Roca: Rio de Janeiro, Brazil, 2015; p. 182.
21. Schötz, S. A phonetic pilot study of vocalisations in three cats. *Proc. Fon.* **2012**, *2012*, 45–48. Available online: https://portal.research.lu.se/portal/files/5449443/3350432.pdf (accessed on 19 April 2019).
22. Rochlitz, I. *The Welfare for Cats*; Springer: Cambridge, UK, 2005; p. 298.
23. Meowsic. Available online: http://vr.humlab.lu.se/projects/meowsic/catvoc.html (accessed on 19 April 2018).
24. McComb, K.; Taylor, A.M.; Wilson, C.; Charlton, B.D. The cry embedded within the purr. *Curr. Biol.* **2009**, *19*, R507–R508. [CrossRef] [PubMed]

© 2019 by the authors. Licensee MDPI, Basel, Switzerland. This article is an open access article distributed under the terms and conditions of the Creative Commons Attribution (CC BY) license (http://creativecommons.org/licenses/by/4.0/).

Review

What Is so Positive about Positive Animal Welfare?—A Critical Review of the Literature

Alistair B. Lawrence [1,2,]*, Belinda Vigors [1] and Peter Sandøe [3,4]

1. Scotland's Rural College (SRUC), West Mains Road, Edinburgh EH9 3RG, UK; belinda.vigors@sruc.ac.uk
2. Roslin Institute, University of Edinburgh, Penicuik EH25 9RG, UK
3. Department of Food and Resource Economics, University of Copenhagen, 1958 Frederiksberg C, Denmark; pes@sund.ku.dk
4. Department of Veterinary and Animal Sciences, University of Copenhagen, 1870 Frederiksberg C, Denmark
* Correspondence: alistair.lawrence@sruc.ac.uk

Received: 11 September 2019; Accepted: 7 October 2019; Published: 11 October 2019

Simple Summary: Positive animal welfare (PAW) is thought to have come about as a response to there being too much of a focus on avoiding negatives in animal welfare science. However, despite its development over the last 10 years, it is not clear what it adds to the study of animal welfare. To clarify this, we conduct a review of the literature on PAW. We aim to identify the characteristic features of PAW and to show how PAW connects to the wider literature on animal welfare. We find that the PAW literature is characterised by four features: (1) *positive emotions* which highlights the capacity of animals to experience positive emotions; (2) *positive affective engagement* which seeks to create a link between positive emotions and behaviours animals are motivated to engage in; (3) *quality of life* which acts to give PAW a role in defining an appropriate balance of positives over negatives and; (4) *happiness* which brings a full life perspective to PAW. While the first two are already well situated in animal welfare studies the two last points open research agendas about aggregation of different aspects of PAW and how earlier experiences affect animals' ability to have well-rounded lives.

Abstract: It is claimed that positive animal welfare (PAW) developed over the last decade in reaction to animal welfare focusing too much on avoiding negatives. However, it remains unclear what PAW adds to the animal welfare literature and to what extent its ideas are new. Through a critical review of the PAW literature, we aim to separate different aspects of PAW and situate it in relation to the traditional animal welfare literature. We find that the core PAW literature is small (*n* = 10 papers) but links to wider areas of current research interest. The PAW literature is defined by four features: (1) *positive emotions* which is arguably the most widely acknowledged; (2) *positive affective engagement* which serves to functionally link positive emotions to goal-directed behavior; (3) *quality of life* which serves to situate PAW within the context of finding the right balance of positives over negatives; (4) *happiness* which brings a full life perspective to PAW. While the two first points are already part of welfare research going back decades, the two latter points could be linked to more recent research agendas concerning aggregation and how specific events may affect the ability of animals to make the best of their lives.

Keywords: positive animal welfare; critical review; positive emotions; positive affective engagement; quality of life; happiness

1. Introduction

Positive animal welfare (PAW) is often described as a recent idea or concept. The first formal reference to PAW appears to be in Boissy et al. [1] which was followed a year later by the first conceptual development of the concept [2]. PAW has been described as a reaction against an undue focus on negative aspects of welfare and reduction of harms (e.g., [2–4]). PAW has also been linked to criticism of the Five Freedoms as being too focused on negatives and harms [3,5,6]. However, it is also clear that PAW emerged from wider animal-welfare thinking which was becoming increasingly interested in positive aspects of welfare including positive emotions [1] and providing animals with resources to facilitate positive welfare [7] (see also [2]).

Given that approximately a decade has passed since the PAW concept emerged this seems an appropriate time to better understand exactly what is being added by PAW and to what extent and how the PAW concept links to the wider animal welfare literature. In this paper we will analyse the existing PAW literature to define the key features of PAW coming out of this, and their relationship with the wider animal-welfare literature. For this we have critically reviewed the literature identifying both core PAW writings that clearly contribute to development of the PAW concept, and also work in the wider literature representing relevant areas of research that are clearly linked to PAW. We have reviewed and analysed the core PAW literature to distill four features that currently define how the PAW concept is understood. We have also reviewed how these features of PAW relate to the wider literature drawing out the extent to which PAW is continuous or distinct from the wider animal welfare literature. Thus we aim to make clearer what research areas have contributed to the development of PAW and what PAW uniquely contributes both scientifically and more widely to the debate over animal welfare.

2. Materials and Methods

To review the literature on PAW we carried out a search using Scopus including published work up to 1 March 2019, with the terms "positive welfare" and "animal" in the authors' keywords. This resulted in a total of 12 papers including book chapters. We then repeated the search using the same terms but now including the works' title, abstract and author's keywords. This added an additional 59 works. We then reviewed these 71 works for relevance ending with a total of 38. Of these 38 we found that a number of works made only passing reference to PAW sometimes with vague or no definitions of what was meant by the term. Therefore, we excluded these works thus leaving what we refer to as the core PAW literature which came to 10 papers and book chapters (Table 1a). Our search and selection strategy results in 5 of these 10 core PAW papers being the work of Mellor and colleagues.

We have also included in Table 1b a list of what we refer to as key linking papers. To be included in this list, a paper had to provide a substantial link between one of the 4 features of PAW that we identified from the core PAW literature and the wider scientific literature. Work in this list did not need to refer to PAW or to develop the PAW concept. We did not seek to provide a comprehensive list of such linking papers and in that sense we accept that our choice can be seen as subjective. However, we would argue that it is beyond the purpose of this paper to provide comprehensive literature reviews for the wider literature that relates to the 4 PAW elements we have identified.

Table 1. (**a**) A chronologically organized list of the core positive animal welfare (PAW) literature; (**b**) a list of literature that links to PAW organized chronologically under the 4 features: positive emotions; positive affective engagement; quality of Life; happiness. Explanations for the inclusion criteria for these lists are contained in the Table. The lettering for this table is used in Figure 1 and cross-referenced against the main reference list in Section 3.5.

(**a**). Core PAW literature: Work that specifically refers to positive animal welfare and clearly contributes to the development of the concept (in chronological order):

A. Yeates, J.W.; Main, D.C. Assessment of positive welfare: A review. *Vet. J.* **2008**, *175*, 293–300.
B. Mellor, D.J. Animal emotions, behaviour and the promotion of positive welfare states. *N. Z. Vet. J.* **2012**, *60*, 1–8.
C. Edgar, J.; Mullan, S.; Pritchard, J.; Mcfarlane, U.; Main, D. Towards a 'good life' for farm animals: Development of a resource tier framework to achieve positive welfare for laying hens. *Animals* **2013**, *3*, 584–605.
D. Boissy, A.; Erhard, H.W. How studying interactions between animal emotions, cognition, and personality can contribute to improve farm animal welfare. In *Genetics and the Behavior of Domestic Animal*; Academic Press: London, UK, 2014; pp. 81–113.
E. Mellor, D.J. Enhancing animal welfare by creating opportunities for positive affective engagement. *N. Z. Vet. J.* **2015**, *63*, 3–8.
F. Mellor, D.J. Positive animal welfare states and encouraging environment-focused and animal-to-animal interactive behaviours. *N. Z. Vet. J.* **2015**, *63*, 9–16.
G. Mellor, D.J. Positive animal welfare states and reference standards for welfare assessment. *N. Z. Vet. J.* **2015**, *63*, 17–23.
H. Mellor, D.J.; Beausoleil, N.J. Extending the 'Five Domains' model for animal welfare assessment to incorporate positive welfare states. *Anim. Welf.* **2015**, *24*, 241–253.
I. Krebs, B.; Marrin, D.; Phelps, A.; Krol, L.; Watters, J. Managing aged animals in zoos to promote positive welfare: A review and future directions. *Animals* **2018**, *8*, 116.
J. Lawrence, A.B.; Newberry, R.C.; Špinka, M. Positive welfare: What does it add to the debate over pig welfare? In *Advances in Pig Welfare*; Woodhead Publishing: Duxford, UK, 2018; pp. 415–444.

(**b**). Literature that clearly links PAW to the general animal welfare literature and also to wider science. To be included here papers or reports were required to link substantially and clearly to one of the 4 elements of PAW which we have identified. Work was not required to refer to PAW or to contribute to the PAW concept. We did not seek to provide a comprehensive list of relevant work but to reference what in our judgement were up to five key linking papers for each of the elements (in chronological order):

1. Positive emotions:

    (a) Boissy, A.; Manteuffel, G.; Jensen, M.B.; Moe, R.O.; Spruijt, B.; Keeling, L.J.; Winckler, C.; Forkman, B.; Dimitrov, I.; Langbein, J.; et al. Assessment of positive emotions in animals to improve their welfare. *Physiol. Behav.* **2007**, *92*, 375–397.
    (b) Mendl, M.; Burman, O.H.; Paul, E.S. An integrative and functional framework for the study of animal emotion and mood. *Proc. R. Soc. B Biol. Sci.* **2010**, *277*, 2895–2904.
    (c) Burgdorf, J.; Panksepp, J. The neurobiology of positive emotions. *Neurosci. Biobehav. Rev.* **2006**, *30*, 173–187.
    (d) Berridge, K.C.; Kringelbach, M.L. Pleasure systems in the brain. *Neuron* **2015**, *86*, 646–664.

2. Positive affective engagement:

    (e) Fraser, D.; Duncan, I.J. 'Pleasures', 'pains' and animal welfare: Toward a natural history of affect. *Anim. Welf.* **1998**, *7*, 383–396.
    (f) Panksepp, J. Affective consciousness: Core emotional feelings in animals and humans. *Conscious. Cognit.* **2005**, *14*, 30–80.
    (g) Bracke, M.B.; Hopster, H. Assessing the importance of natural behavior for animal welfare. *J. Agric. Environ. Ethics* **2006**, *19*, 77–89.
    (h) Franks, B.; Higgins, E.T. Effectiveness in humans and other animals: A common basis for well-being and welfare. In *Advances in Experimental Social Psychology*; Academic Press: San Diego, CA, USA, 2012; pp. 285–346.
    (i) Špinka, M.; Wemelsfelder, F. Environmental challenge and animal agency. In *Animal Welfare*; Appleby, M., Mench, J., Olsson, A., Hughes, B.O., Eds.; CABI: Wallingford, UK, **2011**, 27–43.

3. Quality of life:

    (j) Farm Animal Welfare Council. *Farm Animal Welfare in Great Britain: Past, Present and Future*; FAWC: London, UK, 2009; pp. 1–70.
    (k) McMillan, F.D. Quality of life in animals. Views: Forum. *JAVMA* **2000**, *216*, 1904–1910.
    (l) Yeates, J. Quality of life and animal behaviour. *Appl. Anim. Behav. Sci.* **2016**, *181*, 19–26.
    (m) Vøls, K.K.; Heden, M.A.; Kristensen, A.T.; Sandøe, P. Quality of life assessment in dogs and cats receiving chemotherapy—A review of current methods. *Vet. Comp. Oncol.* **2017**, *15*, 684–691.

4. Happiness:

    (n) King, J.E.; Landau, V.I. Can chimpanzee (*Pan troglodytes*) happiness be estimated by human raters? *J. Res. Personal.* **2003**, *37*, 1–15.
    (o) Seligman, M.E.; Steen, T.A.; Park, N.; Peterson, C. Positive psychology progress: Empirical validation of interventions. *Am. Psychol.* **2005**, *60*, 410.
    (p) Webb, L.E.; Veenhoven, R.R.; Harfeld, J.L.J.; Jensen, M.B. What is animal happiness? *Ann. N. Y. Acad. Sci.* **2018**, doi:10.1111/nyas.13983.

3. The Defining Features of Positive Animal Welfare (PAW)

Our search for papers that reference "positive welfare" and "animal" which also clearly contribute to the development of the concept reveals a small core PAW literature of 10 papers and book chapters (Table 1a). Our subsequent content analysis of this work revealed 4 key defining features—positive emotions, positive affective engagement, quality of life and happiness—that we argue define PAW as currently discussed in the literature. In the following sections we review each of these in turn and how they relate to the wider animal welfare literature.

3.1. Positive Emotions

In line with Boissy et al. and de Vere and Kuczaj [1,8], we will use the term emotion as an overarching term to cover subjective experiences in animals, mainly because we believe that in animals the distinctions between terms such as emotions, affect, feelings and subjective experiences are somewhat arbitrary and not currently open to empirical testing.

The core literature is clear that one of the defining features of the PAW concept is that animals have the capacity for experiencing positive emotions: e.g.,

"However, preventing negative welfare in animals is not the same as providing them with opportunities to experience positive emotions and positive welfare" ([4] p. 81).

"As with negative aspects of welfare, the opportunity for animals to have positive experiences, which we describe as positive welfare" ([9], p. 586).

"The increasing importance assigned to the affective (i.e., emotional) states that animals may experience and the associated greater emphasis given to the promotion of positive affective states" ([10], p. 1).

The centrality of positive emotions to PAW seems to be linked to the accumulating evidence that animals can experience positive emotions, coming from different scientific areas including animal behaviour, psychology, neuroscience and animal welfare science.

"This paper presents a rationale that may significantly boost the drive to promote positive welfare states in animals ... .based largely, but not exclusively, on an experimentally supported neuropsychological understanding of relationships between emotions and behaviour, an understanding that has not yet been incorporated into animal welfare science thinking" ([10], p. 1).

"There has been a growing interest in the study of emotions in animals over the last few decades, resulting in the emergence of a discipline referred to as Affective Neuroscience (Panksepp, 1998). Scientists have made huge progress in understanding how animals perceive their environment and the feelings prompted by this perception" ([4], p. 83).

The idea that animals can experience emotions is also linked to a growing interest in animal welfare science over how we conceptualise and assess animal emotions; for example the merits of discrete emotions versus dimensional approaches consisting of core affective characteristics (e.g., valence and arousal) [11]. Given the significant drive to use animal-based measures in on-farm welfare assessment there is also considerable interest in the development of reliable and valid approaches to assess positive emotions under practical conditions (e.g., [12]).

The PAW literature has finally seen the use of terms to describe emotions that might be thought to have previously been largely applied to humans only such as pleasure, enjoyment, fun, excitement. As examples:

"Indeed, what use is there in satisfying an animal's vital needs, if the life the animal then lives is devoid of any enjoyment? ([1], p. 298).

"Negative affects are therefore incompatible with the having fun affect" ([13], p. 12).

"By analogy with human beings, emotional experiences of environmentally engaged aliveness, positive excitement, even euphoria, are considered likely to attend the operation of the 'Seeking' system" ([10], p. 4).

In summary, reference to positive emotions is a key defining feature of the PAW literature; both in terms of emphasising the capacity of animals to experience positive emotions and through the use of terms associated with positive emotion. Potentially, the use of more positive emotion terms in the PAW literature may be reflective of PAW being a reaction to an over-focus on negative aspects of welfare in the wider animal welfare literature. However, as we will describe in the following section, positive emotion in the context of animal welfare is not unique to the PAW litearature.

Positive Emotions: PAW Contrasted to the Wider Literature

Animal emotions have long been seen as a key aspect of animal welfare (e.g., [14]) and the focus on emotions as an element of PAW is consistent with the wider animal welfare literature. However, the emphasis on and the call for positive emotions, seen in the PAW literature, disguises the fact that there has been recognition of positive emotions in animals from the earliest origins of animal welfare concerns; for example in the Brambell Report:

"We accept that animals can experience emotions such as rage, fear, apprehension, frustration and pleasure, though they do display different degrees and types of intelligence which may affect the reaction to particular stress-causing circumstances." ([15], p. 10).

The recognition of positive emotions can also be found in scientific writing that predates PAW (e.g., [16]) and indeed many of the approaches that are proposed to assess positive emotions are the same as the methods previously proposed to measure negative emotions. For example in preference testing, whilst avoidance of a stimulus can be taken as an indication of a negative emotional response, approach behaviour can be taken to indicate a positive emotional response (e.g., [17]). In judgement or cognitive bias testing the animals' response to ambiguous stimuli can be taken to indicate either negative or positive emotional states (e.g., [18]). Similarly in qualitative behavioural assessment (QBA) the method derives assessments of animals' emotional (subjective) state based on human observer assessments that range from negative to positive (e.g., [19]). Play behaviour which is often proposed as a measure of positive emotional state [1] can, in some instances, be argued to result in part from negative states [20]. Perhaps the least ambiguous of all proposed measures of positive emotional state are the so-called '50 kHz' ultrasonic vocalisations (USVs) produced by rats, but even here the form of these USVs may reflect the balance of positive and negative emotions being experienced [21]. In summary, all current approaches proposed to assess positive emotions in animals emphasise the continuity between PAW and the wider welfare literature.

In addition to the idea of positive emotions predating PAW, any claims of a seminal breakthrough in understanding the nature of positive emotions in animals seem to us to be exaggerated. There remains considerable scientific uncertainty about emotions in animals and, perhaps, particularly positive emotions given the relative lack of research on these [4]. For example we lack a 'gold standard' against which to assess and validate emotional state (negative or positive) in animals. It has been suggested that neuroscience provides the validatory evidence for positive emotional states in animals [10,22]. However, this is to ignore that there remains considerable debate within neuroscience itself over the nature of animal emotions including whether animals consciously experience emotions (e.g., [23]) and the typology best used in emotional theory [24]. For us, neuroscience provides interesting and corroborating evidence for positive emotional states in animals but it falls short of being the gold standard validating evidence. Finally, as with the assessment of emotions in general there is a risk of circularity in the assessment of positive emotions, for example where animals are exposed to a putatively positive context and the animals' responses are then labelled as indicators of a positive state without independent validation (e.g., [25]).

As noted above, the PAW literature has seen the use of terms such as pleasure, fun and enjoyment to describe emotional states in animals that might be thought to have been largely restricted to humans. Our analysis of the use of these terms in the PAW literature is that they appear sporadically and are largely undefined. It is also not the case that terms such as pleasure and fun are only to be found in the PAW literature suggesting that PAW is not the essential trigger to the use of these terms. Darwin's famous early account of the emotional expressions of man and animals [26] makes several references to joy and pleasure in animals:

"Under a transport of joy or of vivid pleasure we see this in the bounding and barking of a dog when going out to walk with his master; and in the frisking of a horse when turned out into an open field" ([26], p. 76).

Much later, and following the development of animal behaviour science, scientific work on play in animals has often referred to the 'enjoyment' associated with the behaviour. For example:

"It is difficult, and probably unnecessary, to avoid applying anthropomorphic terms such as 'spontaneous', 'rewarding' and 'enjoyable' to the behaviour of these young animals" ([27], p. 508).

Pleasure has also been used widely to describe a positive emotional state in animals, for example in affective neuroscience:

"In a sense, pleasure can be thought of as evolution's boldest trick, serving to motivate an individual to pursue rewards necessary for fitness" ([28], p. 646).

The term fun, to describe animals' emotions, is also found in the wider literature:

"'Playing' and 'having fun' are almost synonymous" ([29], p. 463).

A recent special issue of the journal *Current Biology* contained a curated set of papers on fun in humans and animals including a covering editorial:

"As usual with an evolutionary question it is helpful to take a broad look at what appear to be similar behaviours in other species—in particular, to consider fun in other animals, and what functions it might have that could contribute to their evolutionary fitness" ([30], R1).

Hence, perhaps contrary to expectations, the PAW literature does not represent a substantial core of writing on positive states such as joy and fun applied to animals. It would seem that PAW is essentially similar to other areas of science where recently it has become more acceptable to apply positive terms such as fun to describe animals' state and an example of convergent evolution across science areas rather than PAW research leading the way.

Whilst we would argue that the centrality of positive emotions with respect to PAW is more of an evolution than a revolution from previous research on negative emotions, we do think it reasonable to suggest that PAW has helped bring greater attention and interest to the study of positive emotions in animals [3]. Arguably, the contribution of the PAW literature here has been to make a more explicit connection between the capacity for animals to experience positive emotions and the implications of this for their welfare.

We think it is interesting to reflect on why positive emotions are seen by some as core to the PAW concept. For example when discussing how to promote PAW, Yeates and Main [2] base their argument almost entirely on the relationships between negative and positive emotions (see p. 297), appearing to suggest that PAW and positive emotional state are more or less synonymous. This suggests that it is the growing focus, scientific and otherwise, on positive emotions that provides the justification for PAW in order to more fully recognise this capacity in animals; in effect the converse of arguing that there has been too much emphasis on negative emotional states. There may also be an influence here of the continuing interest in animal welfare science for assessing emotional states under practical (farm) conditions which is now extending to positive emotional states. It may also be that scientists are drawn to the study of positive emotions either because they themselves find reciprocal pleasure in studying animals in positive welfare states (e.g., [27]) or because they recognise the advantage of working on a subject ('happy animals') that the public finds engaging (e.g., [31]); for example the findings of a recent paper on the ability of goats to distinguish between positively and negatively valenced calls [32] were 'tweeted' 118 times and mentioned in 128 news stories [33].

Whatever the reason for the focus on positive emotions in the context of PAW we would point out the obvious, that our acceptance that animals can experience positive emotions does beg the question of how these emotions emerge in the first place and what their role is. As we present in the following section, a further element of the PAW literature—positive affective engagement—seeks to make such a functional link between goal-directed behaviours and positive emotions.

3.2. Positive Affective Engagement (PAE)

The need to explain how positive emotions emerge has also been a major defining feature of the PAW literature. One interpretation for the function of emotions is that they are closely linked to goal-directed behaviours. Mellor has written extensively about this in the context of PAW and coined the term positive affective engagement (PAE) which he defines as:

> "... the experience animals may have when they actively respond to motivations to engage in rewarding behaviours, and it incorporates all associated affects that are positive" ([22] p. 3).

PAE, as defined by Mellor [22], proposes that animal emotions are closely, if not intrinsically, linked to the guiding of goal-directed behaviours in what he refers to as the 'emotion-motivation nexus'. In Mellor's proposal, stimuli which are perceived as positive, produce a state of pleasure that acts as a reward and as such reinforces subsequent behaviour. The relevance of PAE for PAW is that it provides a functional link between behaviours such as foraging, maternal care and play and positive subjective experiences. In other words, it anchors positive emotions to functionally important behaviours (see also other related papers by Mellor [13,31]).

PAE, as developed by Mellor, links closely to affective neuroscience which studies how the brain generates subjective sensations including those of reward and pleasure. For example Mellor [10] uses the writings of Panksepp (e.g., [23]) who suggests that the brain has a number of discrete 'action-orientated systems' (including SEEKING and PLAY) that integrate emotions and motivation to organise specific behavioural responses. Mellor (e.g., [22]) also refers to the work of Berridge and co-workers (e.g., [28,34]) who have researched the neural substrates for 'wanting' (motivations) and 'liking' (sensations of pleasure) that are integral to reward systems (see also Yeates and Main [2] who discuss wanting and liking at length). Again, whilst this neuroscience research provides some corroboration for PAE we consider that there is still a considerable gap between neuroscience experimental models and the complex spontaneous behaviours (e.g., foraging, exploration, play) which are the focus of PAE.

PAE also links to other ideas on behavioural expression that have relevance to PAW. For example, Mellor makes reference to the importance of animals being able to "exercise of voluntary, self-generated behavioural expression" ([22] quoting [35]). Edgar et al. [9] base their resource-tier approach on providing animals with what they refer to as 'good life opportunities' specifically to allow animals the choice to engage in behaviours that will elicit positive emotions. The good life opportunities they list (comfort, pleasure, interest, confidence and health) are mainly taken from the UK's FAWC report on a future vision for farm animal welfare [5]. In sum, PAE highlights the interconnection between behavioural expression and positive emotions and arguably situates this within PAW to draw attention to the welfare-relevance of behaviours which promote positive experiences in animals. However, as we present in the following section, the wider animal welfare literature also considers the functional link between behaviours and emotion.

Positive Affective Engagement: PAW Contrasted to the Wider Literature

We again see considerable continuity and overlap between PAE, with its emphasis on the intrinsic links between positive emotions, motivations and goal-directed behaviours, and the wider animal welfare literature. For example Fraser and Duncan [16] proposed the very similar concept of motivational affective states (MAS) as adaptations where subjective states are involved in motivating specific behaviours. Interestingly, Fraser and Duncan [16] make a case for negative and positive MAS serving different functions. Negative MAS (e.g., hunger; thirst) have the function of resolving needs, whereas positive MAS (e.g., play or exploration) occur to exploit opportunities when other more pressing needs are not present. The idea that negative and positive MAS might serve different functions in animals, is interestingly similar to the more recent human-based 'broaden and build' theory [36] which proposes that positive emotions may facilitate broadening of the mindset through play and exploration (once basic needs are met).

Other writings also propose a close link between motivation, emotions and welfare. Franks and Higgins [37] propose that in animals, like humans, well-being will be highest when animals are able to be effective in their pursuit of motivations. The importance of animals being able to express voluntary behavior [35] has been developed into the concept of animal agency [38–40] defined as 'inner-motivated behavioural engagement with the environment' which can be argued to be a key aspect of positive welfare (e.g., [39]).

In fact the continuity between PAW and the more general animal welfare literature with respect to the proposed close relationship between behavioural expression and welfare can be argued to go as far back as the Brambell Report: e.g.,

"In principle we disapprove of a degree of confinement of an animal which necessarily frustrates most of the major activities' which make up its natural behaviour" ([15], p. 13).

The emphasis on behavioural expression in the Brambell Report (including in T.H. Thorpe's Appendix [15], p. 71) seems to have directly led to FAWC distilling the 'Freedom to express normal behaviour'; the use of *normal* here placing emphasis on the animals' current state rather than its previous (natural) state [41]. As a number of authors have pointed out (e.g., [2]) this freedom is not explicitly about positive welfare and more about preventing a negative (absence of frustration); yet it could equally be argued that this freedom also allows behavioural expression and associated positive emotions [42].

This early emphasis on behavioural expression and welfare was to prove an enduring focus for animal welfare research. The research that followed reasoned that normal behaviour (i.e., behaviour observed under unconstrained conditions) could be regarded as a 'behavioural need' which, therefore, required that housing provide the necessary 'obligatory' features to 'trigger' and satisfy the underlying motivation [43]. Following this early research, behavioural needs entered a phase dominated by analysis of the strength of underlying motivations in order to base housing requirements on the animals own priorities (e.g., the application of 'consumer demand' theory: [44,45]). Here the focus was on identifying the animals' 'necessities' or 'wants' in order to minimise suffering caused by preventing strongly motivated behaviours. As pointed out by Hughes and Duncan [46], one risk of this approach is that the list of necessities as defined by motivational analysis may be insufficient to protect welfare and we would argue that this may be particularly so in the context of promoting PAW. A particular concern with respect to PAW on basing housing design and other management decisions on approaches such as consumer demand is that behaviours often regarded as potential indicators of PAW (e.g., play, exploration, positive social behaviours) may not be resilient in the presence of other stronger motivations such as hunger [47]; in the terminology of consumer demand, behaviours such as play and exploration may come to be defined as 'luxuries'.

We would again argue that PAE, developed in the context of PAW, is a clear continuation of past conceptual work. However, it is also the case that this approach appears to have changed thinking with respect to the relationship between behavioural expression and welfare. This is particularly noticeable in the context of the debate over behavioural needs, where PAE appears to have reversed the argument back towards the original conception of behavioural needs; i.e., housing and management should facilitate normal behaviour with even an emphasis on what might have been regarded as luxury behaviours. A good example of this is the resource-tier approach [9] that aims to provide for behaviours such as curiosity driven exploration and play; behaviours that would likely not have been prioritised using a motivational analysis approach. Such explicit reference to positive experiences and their welfare relevance may be a further outcome of the PAW literature's desire to overcome what has been seen as a narrow focus on negative aspects in animal welfare.

We would also argue that concepts such as PAE and MAS occupy a central space in PAW as they offer a link between positive emotions and the expression of evolutionary adaptive goal-directed behaviours which in turn links through to other relevant ideas and concepts. One of these is quality of life, which attempts to say something about the animals' overall state of welfare, and where the opportunity to perform positively motivated behaviours has been highlighted as a key element of a 'good life' [5,9]. As we discuss in the following section, while positive emotions and PAE have focused on what contributes to the positive aspects of an animal's life, quality of Life brings a perspective on how PAW can be used to characterise the overall welfare state of an animal.

3.3. Quality of Life

The quality of life (QoL) concept was effectively introduced to PAW by Yeates and Main [2] concluding that it is possible to conceive of PAW as a continuum from negative to positive:

"... the idea of a welfare continuum may be a valid model for overall welfare assessment" ([2], p. 297).

As noted already, the basis of their argument is entirely based on scientific understanding of emotions and specifically the relationships between negative and positive emotions:

"Consequently, it does appear possible to describe positive and negative affect in a model that accepts the negative correlation between them, even if this is not perfect" ([2], p. 297).

The FAWC (2009) report [5] followed this by conceiving of animal welfare in terms of a QoL scale, although without specifically referring to PAW:

"Our proposal, therefore, is that an animal's quality of life can be classified as a life not worth living, a life worth living and a good life. Other classification schemes use four or more levels; three have the merit of simplicity and the basic notions are familiar in the human context" ([5], p. 17).

This has proved an influential idea; currently the FAWC report [5] has been cited over 120 times, considerably more than the average for FAWC reports (Google Scholar). It has been partly responsible for triggering re-thinking over the completeness of the 5 freedoms as a welfare framework (e.g., [48]) and led to scientific-based studies based on FAWC's QoL approach such as the resource-tier study [9]. The three-level QoL approach is now mirrored in a number of commercial farm assurance schemes (e.g., [49]).

In summary, the minimal development of QoL within the PAW literature conceives of QoL as a continuum from negative to positive with positive welfare situated at the higher end of the continuum based on either the animals' overall emotional state [2] or the available opportunities for the animal to have a good life [5].

Quality of Life: PAW Contrasted to the Wider Literature

QoL is again not unique to PAW. It was originally developed in human medicine before being translated to veterinary medicine by McMillan [50]. For McMillan [50] animal QoL is defined by emotional experiences:

"Affect (subjective feelings) plays a preeminent and, I propose, exclusive role in all interpretations of QoL in animals." ([50], p. 1905).

McMillan's conception of animal QoL, as defined by emotion, is of a continuum ranging from unpleasant to pleasant experiences, a good QoL being where pleasant experiences outweigh the unpleasant over the lifetime (just exactly what the ratio should be is not defined). Subsequent developments of QoL applied to animal welfare [51] have also emphasised the key relationship between QoL and the animal's subjective experiences. However QoL in humans is seen as a multi-dimensional concept (e.g., [52]) consisting of different domains, suggesting that QoL approaches in animals which are unduly focused on any single aspect or domain may not be inclusive enough to capture accurate assessments of QoL. For example, a recent analysis of QoL studies in dogs and cats receiving chemotherapy [53] found that these studies were mainly focused on assessing the animals' physical state. Based on QoL approaches for infants, Vols et al. [53] suggest that it is possible to assess animal QoL on a wider range of domains including social and role functioning (see also [54,55]).

As we discussed earlier, there has also been debate over the dimensionality of the motivation–emotion nexus; for example Fraser and Duncan [16] suggested that negative and positive MAS may serve different functions and in that sense be orthogonal to each other. Yeates and Main [2] similarly discuss the relationships between negative and positive emotions and conclude in favour of a single continuum, partly on the basis that positive and negative emotions can mutually inhibit each other (e.g., [54]).

This third feature of PAW leads directly into a discussion about aggregation, i.e., about how to add up different aspects of welfare to a total value. The discussion about this has emerged in connection with attempts to define comprehensive measures of animal welfare at farm or flock level and is also mirrored in Mellor and colleagues' development of the 5 domains approach to welfare assessment [55].

Traditionally, animal welfare research has focused on applying single welfare indicators, often in an experimental setting, and the question of how to provide a more comprehensive assessment of an animals' welfare was largely avoided. Since the 1990's this has changed gradually with the development of systems for assessing overall welfare impact in laboratory animals [56,57] and, since around 2000, initiatives developed to assess farm animal welfare at group level. These initiatives have given rise to more systematic discussions about how to integrate different aspects of what matters to animals (e.g., [58,59]). These efforts have so far culminated in the Welfare Quality® project, that developed protocols to measure the welfare of cattle, pigs and hens at farm level (see [60]).

In these initiatives, a growing understanding has developed of the variety of measures which can be considered to give a comprehensive account of the welfare of animals in a given setting. However, there have only been few attempts to explain and justify, firstly, how to add up these measures to give an account of the net welfare of the affected animals and, secondly, how to draw lines between positive, neutral and negative welfare states. So far the most developed attempt to do this in the Welfare Quality® project has attracted severe criticisms of both a conceptual and an ethical nature [61–64].

In sum, QoL arguably serves to further convey the importance of considering more than the negative aspects of an animal's life, highlighting the relevance of assessing the animal's overall welfare state (inclusive of both negative and positive aspects). However, while the two previously discussed features of PAW—positive emotions and PAE—are both linked to a wider and well established literature, the development of QoL as a key feature of PAW will require more development of the link between PAW and the emerging research agenda on QoL. We see this as particularly concerning: how to derive a comprehensive set of measures covering all aspects of animal welfare; how to add up these measures to give a representation of the net welfare; and, how to draw the lines between positive, neutral and negative net welfare. We will now move onto the final defining feature of PAW, happiness in animals, which is distinct from QoL as being a conception of the animals' welfare over its lifetime.

3.4. Happiness in Animals

Happiness was introduced to PAW by Yeates and Main [2]; indeed they open their paper with reference to a poem by Alexander Pope on human happiness. However they went further when they suggested that there might be animal equivalents to categories of human happiness proposed by positive psychology:

> "Human psychologists have advocated dividing the human happiness into (1) 'the pleasant life', (2) 'the engaged life' and (3) 'the meaningful life' [55]. Tentative analogues for animals might be (A) everyday sensational pleasures, (B) engaging with their environment, their conspecifics and their handlers and (C) realising their own goals" ([2], p. 296).

Yeates and Main [2] presented these happiness categories as separate entities, and there is no direct linkage made here between sensations of pleasure and the other categories (e.g., engagement with the environment). However, it is possible to see this idea of happiness as a call for looking at the full life of an animal as a basis for saying that the animal has achieved positive welfare. This contrasts with the literature described in connection with QoL where the set of measures applied to animals typically only aim to look at the state of the animals at a specific point in time. This seems a logical extension of the PAW concept.

Happiness in Animals: PAW Contrasted to the Wider Literature

We found that happiness in animals has been discussed and researched in the wider literature, and that the term is used in different ways. For example, one line of work has developed from the study of subjective well-being (SWB) in humans which is also referred to as happiness (e.g., [65]). In a series of papers starting with King and Landau [66], a SWB rating scale developed for humans was modified for zoo keepers to assess happiness in captive primates; this approach has subsequently been applied to captive primates to study the genetics of happiness [67], lifetime changes in happiness [68], and the relationship between happiness and longevity [69]. More recent work has studied the relationship between SWB (happiness) and welfare, again in primates using zoo keepers to assess these attributes on the same animals using different scales [70]. Other experimental work has defined animal happiness as more optimistic performance in a judgement bias test [71]. More recently Webb et al. [72] have reviewed animal happiness integrating human and animal research and concluded that animal happiness can be conceived of as a long-term relatively stable trait that reflects the balance of negative and positive emotions summarised as 'how an animal feels most of the time'.

The notion of happiness could bring the lived life of each individual animal back into focus when it comes to discussions about PAW, including so far undiscussed aspects such as how experiences across an animal's life (including during early life) may affect PAW by affecting the ability of animals to make the most of the available opportunities (referred to as live-ability by Webb et al. [64]).

In summary, as with QoL the study of animal happiness is much less established than the study of positive emotions and PAE. Here the idea of PAW may help to inspire an emerging research agenda where the focus is on how events across an animals' different life phases must interact for animals to be genuinely happy (i.e., to achieve a well-rounded life).

As we have discussed in the preceding sections, PAW can be characterised by four key features; positive emotions, positive affective engagement, Quality of Life and happiness, which vary in their overlap with the wider animal welfare literature. As such, although we would argue that the core elements of PAW are not distinct from the wider animal welfare literature, what the PAW literature has perhaps done is create a space where these elements are more clearly exemplified and, by doing so, made more manifest the importance of looking at the positive end of the scale in the context of animal welfare.

In the following section, we focus more closely on the interconnections between the core elements of PAW and the wider animal welfare literature.

3.5. Interconnections between Features of PAW and the Wider Literature

Our review of the PAW literature has revealed that, despite consisting of only 10 pieces of work, it is complex with a number of features and links to the wider literature. In order to better understand this complexity we have developed a qualitative interpretation of how the different elements of PAW interrelate and link to the wider literature (Figure 1). The following points explain how we arrived at this interpretation: (see Figure 1 legend for explanation of the notation used): (here in the text we cross-refer to papers from the main reference list as [1–76] with the notation used in Table 1 (A, B, C, D, E, F, G, H, I, J) and (a, b, c, d, e, f, g, h, i, j, k, l, m, n, o, p) (Note: Figure 1 only uses the notation from Table 1):

I. <u>Motivation–emotion</u>: we have placed the motivation–emotion nexus as central to PAW (in effect combining the features of *positive emotions* and *positive affective engagement*) because this resulted in the most optimal arrangement of links between PAW elements and with the wider literature. We also agree with the proposal that animals have evolved a close and intricate relationship between goal-directed behaviour and emotion to better adapt and fulfil their evolutionary goals (e.g., [22] (B)). We would also further propose in a natural state such is the close relationship between motivation and emotion that they can in effect be regarded as a single complex or entity (as proposed by Fraser and Duncan [16] (e)); this approach also avoids debate over the primacy of motivation or emotion ([3] (J)). Note: as indicated in the right hand corner box we have assumed an overlap between the motivation-emotion nexus and the concepts of 'effectiveness in animals' ([37] (h)) and 'animal agency' ([39] (i)).

Moving clockwise around the Figure starting at the bottom with:

II. <u>Adaptations</u>: Mellor (e.g., [22] (B)) and Fraser and Duncan ([16] (e)) both see the neural components underlying motivation and associated emotions as 'genetic pre-adaptations'; this links with both affective neuroscience (e.g., Panksepp ([23] (f)), Berridge and colleagues (e.g., [28] (d)), and also the wider animal welfare literature that has long had an interest in animals living 'natural lives' and being able to 'perform most normal behaviour' (e.g., [42] (g)). Note: both Fraser and Duncan ([16] (e)) and Bracke and Hopster ([42] (g)) make specific reference to the pleasure associated with performance of specific natural (or normal) behaviours.

III. <u>Choice</u>: choice has been another major theme of the wider animal welfare literature (e.g., as a means of assessing 'behavioural needs' [17]) and clearly links to the motivation–emotion nexus. Choice also features in the core PAW literature ([9] (C)) and is part of the rationale for the concept of giving animals 'good life opportunities' (e.g., [5] (j)). As we indicated earlier, one of the features of good life opportunities is that they are aimed at facilitating both necessities (e.g., foraging when hungry) and luxuries (e.g., play behavior) ([9] (C)).

IV. <u>Reward</u>: reward is an important aspect of the core PAW literature with both Yeates and Main ([2] (A) and Mellor (e.g., [22] (B) discussing the implications of the work of Berridge and colleagues (e.g., [28] (d)), which suggests a distinction between the psychological components of reward in terms of their underlying neural substrates. 'Wanting' (or the motivational drive involving incentive salience) is generated by a substantial and distributed brain system. 'Liking' (or the pleasure component of reward) is generated by a small number of 'hedonic hot spots'. Even across very small distances within the same brain centre, different behavioural responses indicative of wanting and liking can be generated (e.g., [28] (d)). This work draws a potential distinction between motivation and emotional responses to stimuli in terms of how the brain generates these sensations; however, as pointed out by Yeates and Main ([2] (A)), this distinction may rarely be relevant under 'normal' conditions. Note: there seems to be a clear link between the neuropsychology literature on wanting and the motivationally based use of choice and opportunities in PAW.

V. <u>Happiness</u>: happiness is one of the defining features of PAW and was introduced by Yeates and Main ([2] (A)) in referring to the work of human based positive psychologists such as

Seligman ([55] (o)). Happiness in animals has also been reviewed recently by Webb et al. ([72] (p)) somewhat separately to PAW. Human happiness is often distinguished into the categories of pleasure (hedonic happiness) and meaning or purpose (eudaimonic happiness); Yeates and Main ([2] (A)) consider what might be animal equivalents to these separate human-based happiness categories. Note: We have introduced the term 'doing' as a more relevant animal-based term to cover eudaimonic happiness. We propose that the pleasure aspects of happiness, the liking component of reward and the pleasure associated with performing normal behavior are all strongly overlapping. Webb et al. ([72] (p)) introduce the concept of affective happiness which can be thought of as how an animal feels most of the time and this we propose overlaps with this conception of happiness and the application of the QoL concept to PAW. We propose that happiness is distinct from QoL, by drawing attention to the whole of the lived life of the animal and how experiences at different life stages can influence how much the animal is able to have a rounded and positive life.

VI. Quality of life: *QoL* is another defining PAW feature largely because of the FAWC (2009) report ([5] (j)) that introduces the idea of a 'good life' (as the upper band of a QoL scale for farm animals). QoL is also discussed in the context of PAW by Yeates and Main ([2] (A)) and Mellor ([73] (H)). However we found QoL in the context of PAW to be the most confusing element. FAWC ([5] (j)) do not specifically refer to PAW although it is reasonable to assume that a good life refers to a PAW state. QoL applied to PAW appears to be largely based on the animal's emotional state (see both ([2] (A)) and ([73] (H))) whereas QoL is more usually seen as consisting of domains (one of which relates to emotional state) (e.g., [53] (m)). Furthermore within PAW there is the development of the Five Domains model of welfare assessment ([73](H)) which for us could also be described as a QoL assessment approach. Therefore, we see a need to develop QoL in the context of PAW by developing approaches to allow different welfare aspects (negative and positive) to be integrated and to place an animals' welfare on a scale from bad to excellent.

VII. The study of affect (emotions) in animals: this is a substantial field of work that we only refer to through key linking references. We distinguish between research on theories and concepts of affect/emotions in animals (e.g., [11] (b)) and the assessment of affect/emotions in animals (e.g., [1] (a)). Interestingly the most direct referencing to PAW can be found in the assessment literature and especially ([1] (a)) which makes several references to the motivation–emotion nexus.

VIII. Cognition: This is the first time we have specifically referred to cognition as we found it not to be a substantial defining feature of PAW. It is referred to in passing by Mellor (e.g., [22] (E)) and by ([74] (D)) and ([75] (I)) in the context of environmental enrichment strategies.

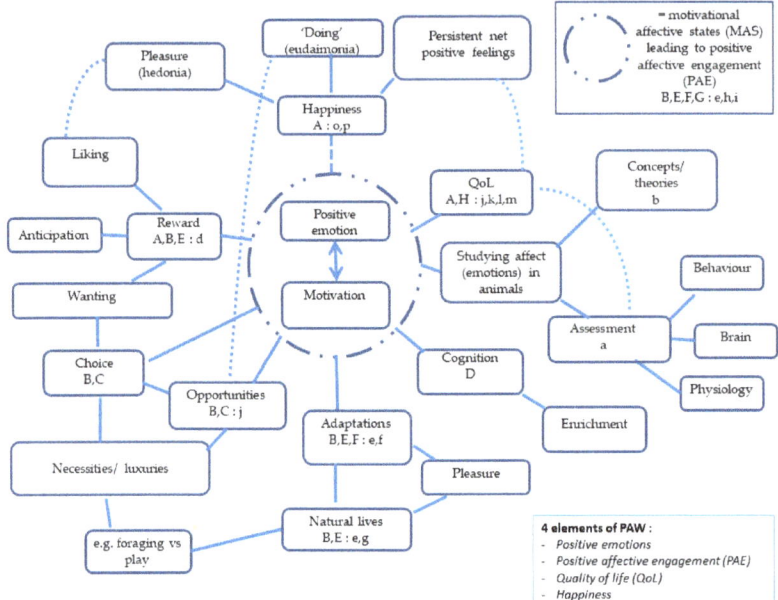

Figure 1. *An interpretation of the inter-relationships between the defining elements of PAW and the wider literature.* Reference lettering is taken from Table 1 and distinguishes between the core PAW and wider literature. Top right box: provides references to the motivation–emotion nexus; Bottom right box: lists the 4 defining elements. Solid lines between terms indicates a definable link has been made in the literature; dotted lines propose potential links/overlaps. We have not included the dotted line linking the pleasure from performing natural behavior to liking and happiness pleasure as this interferes with the readability of the figure.

4. Discussion

We undertook this review of PAW to better understand how the PAW concept has developed since it was first introduced and also to critically assess how PAW interrelates with and adds to the wider animal welfare literature. Our review shows clearly how small the specific PAW literature is (defined as those papers and book chapters that refer to PAW and clearly develop the concept); it is also the case that of the 10 papers that we identified as core PAW papers, 5 of these are the work of Mellor and colleagues. At the same time, our review illustrates how relatively complex PAW is, being constituted of a number of concepts and ideas emerging from different fields. We found the 4 features of positive emotions, positive affective engagement (PAE), quality of life (QoL) and happiness to define the current PAW literature. We found little evidence to suggest that PAW represents a 'step-change' in thinking but is better seen as an evolution from the wider animal welfare literature. Indeed the distinction between the PAW literature and wider animal welfare literature is somewhat arbitrary; in our case being dependent on our search methodology and whether the paper specifically referred to PAW in the title, abstract or author's keywords. As a result we distinguish between the core PAW literature and papers that best linked the PAW literature to wider, relevant areas of research that did not specifically refer to positive welfare (Table 1).

Among the defining features, we noted the centrality of positive emotions to the development of PAW and discussed various possibilities for why this is, including that for some the growing evidence for positive emotions in animals provides the necessary justification for PAW (e.g., [1,2]). Despite the logic of this position we also see risks in too closely defining PAW through positive emotions including that such a focus reduces attention to other important inputs to overall welfare; indeed it begs the question of how positive emotions emerge in the first place. In our synthesis of the interrelationships between PAW and related concepts (see Figure 1), we propose the emotion–motivation nexus (PAE, in brief) as central because it provides both a link between goal-directed behaviours and the emergence of positive emotions, and also provides links between PAW and the wider literature. As examples, PAE clearly overlaps with the happiness components of doing and the resulting experiencing of pleasure; PAE also links to QoL through the FAWC proposal to provide animals with good life opportunities, which effectively is about providing animals with opportunities for PAE [5,9]. We acknowledge that there may be other ways of visualising these interconnections.

Our review has clearly identified the significant overlap between the concepts and ideas that have variously contributed to PAW (see also Figure 1). PAE, for example, overlaps with existing writings on performing natural (or normal) behaviours (e.g., [42]), motivational affective states (MAS) [16], effectiveness [37] and animal agency (e.g., [38]), all of which point to the close relationships between initiating and completing goal-directed behaviours and positive emotions. PAE also overlaps with ideas of giving animals choices and providing good life opportunities with which to express 'positive' goal-directed behaviours [9], and with the study of reward in animals which attempts to disaggregate the motivational and positive emotional aspects of reward seeking behavior (e.g., [28]). As we propose above there are overlaps between the motivation–emotion nexus and happiness, given that happiness in animals can be seen as involving the achieving of goals (doing) and the resulting pleasure (feeling good) [2].

In our view the overlaps between various ideas about how PAW relates to longer-term welfare are less clear because here PAW interacts with only recently emerging research agendas in the study of animal welfare focusing on how to aggregate welfare both at a specific moment of time and across and individual animal's life (e.g., [61]). In this context the idea of animal happiness adds the important time dimension to PAW but otherwise seems closely overlapping with QoL. Indeed happiness and QoL could be combined if we accept that for the individual animal, it is what is experienced over the longer-term which sums up the negatives and positives of life (see also [55]). Integrating happiness with QoL in this way would bring the lived life of each individual animal into focus when it comes to discussions about PAW. As we discussed earlier, an important discussion here is about how animals' experiences during different life phases may affect their future PAW or animals' ability to make the best out of the opportunities they are given (see also [72]).

We have argued that the 4 defining features of the PAW literature are to varying degrees extensions of the wider animal welfare literature and in that sense PAW seems to be a natural development of animal welfare rather than a 'step-change'. However, it also seems reasonable to us to propose that PAW could be an effective route to changing attitudes to animals and to farming [3], particularly if clear consideration is first given to how key stakeholders in society interpret PAW and how this can affect their response to it [76].

5. Conclusions

Our analysis of the PAW literature suggests that it can be defined by 4 features (positive emotions, positive affective engagement (PAE), Quality of Life (QoL) and happiness). The first two features (positive emotions and PAE (or the reward from completing complex motivated behaviours)) are already part of welfare research going back decades. However, the two latter features (QoL and happiness) link to more recent research agendas concerning how we integrate or add up a wide range of welfare measures, and how specific life events may affect the ability of animals to make the most of their life opportunities. Overall, similar to the arguments that are made over positive human psychology, PAW emphasises the capacity for animals to live good lives which in turn could inspire higher aspirations for animal welfare standards.

Author Contributions: A.B.L. and P.S. conceived the study. A.B.L. drafted the manuscript. P.S. and B.V. contributed to the interpretation of the literature review and the writing and proof-reading of the manuscript.

Funding: This research was supported by funding from the Scottish Government's Rural and Environment Science and Analytical Services Division (RESAS). Alistair Lawrence also receives support from the Roslin Institute (University of Edinburgh) BBSRC Institute Strategic Programme.

Acknowledgments: The authors acknowledge useful inputs through discussions from various colleagues including Linda Keeling and David Mellor.

Conflicts of Interest: The authors declare no conflicts of interest.

References

1. Boissy, A.; Manteuffel, G.; Jensen, M.B.; Moe, R.O.; Spruijt, B.; Keeling, L.J.; Winckler, C.; Forkman, B.; Dimitrov, I.; Langbein, J.; et al. Assessment of positive emotions in animals to improve their welfare. *Phys. Behav.* **2007**, *92*, 375–397. [CrossRef] [PubMed]
2. Yeates, J.W.; Main, D.C.J. Assessment of positive welfare: A review. *Vet. J.* **2008**, *175*, 293–300. [CrossRef] [PubMed]
3. Lawrence, A.B.; Newberry, R.C.; Špinka, M. Positive welfare: What does it add to the debate over pig welfare? In *Advances in Pig Welfare*; Špinka, M., Ed.; Herd and Flock Welfare; Woodhead Publishing: Duxford, UK, 2018; pp. 415–444.
4. Phillips, C. Animal welfare: A construct of positive and negative affect? *Vet. J.* **2008**, *175*, 291–292. [CrossRef] [PubMed]
5. FAWC. *Farm Animal Welfare in Great Britain Past Present and Future*; Farm Animal Welfare Council: Nobel House, London, UK, 2009; pp. 1–70.
6. Mellor, D.J. Moving beyond the "Five Freedoms" by Updating the "Five Provisions" and Introducing Aligned "Animal Welfare Aims". *Animals* **2016**, *6*, 59. [CrossRef] [PubMed]
7. Morton, D. A hypothetical strategy for the objective evaluation of animal well-being and quality of life using a dog model. *Anim. Welf.* **2007**, *16*, 75–81.
8. De Vere, A.J.; Kuczaj, S.A. Where are we in the study of animal emotions? *Wiley Interdiscip. Rev. Cogn. Sci.* **2016**, *7*, 354–362. [CrossRef]
9. Edgar, J.L.; Mullan, S.M.; Pritchard, J.C.; McFarlane, U.J.C.; Main, D.C.J. Towards a 'Good Life' for Farm Animals: Development of a Resource Tier Framework to Achieve Positive Welfare for Laying Hens. *Animals* **2013**, *3*, 584–605. [CrossRef]
10. Mellor, D. Animal emotions, behaviour and the promotion of positive welfare states. *N. Z. Vet. J.* **2012**, *60*, 1–8. [CrossRef]
11. Mendl, M.; Burman, O.H.P.; Paul, E.S. An integrative and functional framework for the study of animal emotion and mood. *Proc. R. Soc. B Biol. Sci.* **2010**, *277*, 2895–2904. [CrossRef]
12. Napolitano, F.; Knierim, U.; Grass, F.; De Rosa, G. Positive indicators of cattle welfare and their applicability to on-farm protocols. *Ital. J. Anim. Sci.* **2009**, *8*, 355–365. [CrossRef]
13. Mellor, D.J. Positive animal welfare states and encouraging environment-focused and animal-to-animal interactive behaviours. *N. Z. Vet. J.* **2015**, *63*, 9–16. [CrossRef] [PubMed]
14. Duncan, I.J.H. Welfare is to Do with What Animals Feel. *J. Agric. Environ. Ethics* **1993**, *6*, 8–14.
15. Brambell, F.W.R. *Report of the Technical Committee to Enquire into the Welfare of Animals kept under Intensive Livestock Husbandry Systems*; Her Majesty's Stationery Office: London, UK, 1965.
16. Fraser, D.; Duncan, I.J.H. "Pleasures",'Pains' and Animal Welfare: Toward a Natural History of Affect. *Anim. Welf.* **1998**, *7*, 383–396.
17. Dawkins, M. Do hens suffer in battery cages? environmental preferences and welfare. *Anim. Behav.* **1977**, *25*, 1034–1046. [CrossRef]
18. Burman, O.H.P.; Parker, R.; Paul, E.S.; Mendl, M. A spatial judgement task to determine background emotional state in laboratory rats, Rattus norvegicus. *Anim. Behav.* **2008**, *76*, 801–809. [CrossRef]
19. Wemelsfelder, F.; Hunter, A.E.; Paul, E.S.; Lawrence, A.B. Assessing pig body language: Agreement and consistency between pig farmers, veterinarians, and animal activists1. *J. Anim. Sci.* **2012**, *90*, 3652–3665. [CrossRef] [PubMed]
20. Ahloy-Dallaire, J.; Espinosa, J.; Mason, G. Play and optimal welfare: Does play indicate the presence of positive affective states? *Behav. Process.* **2018**, *156*, 3–15. [CrossRef]

21. Barker, D.J. Ultrasonic Vocalizations as an Index of Positive Emotional State. In *Handbook of Behavioral Neuroscience*; Brudzynski, S.M., Ed.; Handbook of Ultrasonic Vocalization; Elsevier: London, UK, 2018; Volume 25, pp. 253–260.
22. Mellor, D.J. Enhancing animal welfare by creating opportunities for positive affective engagement. *N. Z. Vet. J.* 2015, *63*, 3–8. [CrossRef]
23. Panksepp, J. Affective consciousness: Core emotional feelings in animals and humans. *Conscious. Cogn.* 2005, *14*, 30–80. [CrossRef]
24. Watt, D.F. Panksepp's common sense view of affective neuroscience is not the commonsense view in large areas of neuroscience. *Conscious. Cogn.* 2005, *14*, 81–88. [CrossRef]
25. Clarke, T.; Pluske, J.R.; Collins, T.; Miller, D.W.; Fleming, P.A. A quantitative and qualitative approach to the assessment of behaviour of sows upon mixing into group pens with or without a partition. *Anim. Prod. Sci.* 2017, *57*, 1916–1923. [CrossRef]
26. Darwin, C. The Expression of the Emotions in Man and Animals. Available online: http://darwin-online.org.uk/content/frameset?itemID=F1142&viewtype=text&pageseq=1 (accessed on 5 August 2019).
27. Martin, P.; Bateson, P. The ontogeny of locomotor play behaviour in the domestic cat. *Anim. Behav.* 1985, *33*, 502–510. [CrossRef]
28. Berridge, K.C.; Kringelbach, M.L. Pleasure systems in the brain. *Neuron* 2015, *86*, 646–664. [CrossRef] [PubMed]
29. Trezza, V.; Baarendse, P.J.J.; Vanderschuren, L.J.M.J. The pleasures of play: Pharmacological insights into social reward mechanisms. *Trends Pharmacol. Sci.* 2010, *31*, 463–469. [CrossRef] [PubMed]
30. North, G. The Biology of Fun and the Fun of Biology. *Curr. Biol.* 2015, *25*, R1–R2. [CrossRef]
31. Nelson, X.J.; Fijn, N. The use of visual media as a tool for investigating animal behaviour. *Anim. Behav.* 2013, *85*, 525–536. [CrossRef]
32. Baciadonna, L.; Briefer, E.F.; Favaro, L.; McElligott, A.G. Goats distinguish between positive and negative emotion-linked vocalisations. *Front. Zool.* 2019, *16*, 25. [CrossRef]
33. Altmetric Overview of Attention for "Goats Distinguish Between Positive and Negative Emotion-Linked Vocalisations". Available online: https://www.altmetric.com/details/63354813 (accessed on 8 August 2019).
34. Berridge, K.C.; Robinson, T.E. Parsing reward. *Trends Neurosci.* 2003, *26*, 507–513. [CrossRef]
35. Wemelsfelder, F. The scientific validity of subjective concepts in models of animal welfare. *Appl. Anim. Behav. Sci.* 1997, *53*, 75–88. [CrossRef]
36. Fredrickson, B.L. The broaden-and-build theory of positive emotions. *Philos. Trans. R. Soc. Lond. B Biol. Sci.* 2004, *359*, 1367–1378. [CrossRef]
37. Franks, B.; Higgins, E.T. Effectiveness in Humans and Other Animals: A Common Basis for Well-being and Welfare. In *Advances in Experimental Social Psychology*; Zanna, M.P., Olson, J.M., Eds.; Academic Press: San Diego, CA, USA, 2012; pp. 285–346.
38. Wemelsfelder, F.; Birke, L.I.A. Environmental Challenge. In *Animal Welfare*; Appleby, M., Hughes, B.O., Eds.; CABI: Wallingford, UK, 1997; pp. 35–47.
39. Špinka, M.; Wemelsfelder, F. Environmental challenge and animal agency. In *Animal Welfare*; Appleby, M., Mench, J., Olsson, A., Hughes, B.O., Eds.; CABI: Wallingford, UK, 2011; pp. 27–43.
40. Špinka, M. Animal agency, animal awareness and animal welfare. *Univ. Fed. Anim. Welf.* 2019, *28*, 11–20. [CrossRef]
41. Lawrence, A.B.; Vigors, B. Farm animal welfare: Origins and interplay with economics and policy. In *Economics of Farm Animal Welfare: Theory, Evidence and Policy*; CABI: Wallingford, UK, in press.
42. Bracke, M.B.M.; Hopster, H. Assessing the Importance of Natural Behavior for Animal Welfare. *J. Agric. Environ. Ethics* 2006, *19*, 77–89. [CrossRef]
43. Stolba, A.; Wood-Gush, D.G. The identification of behavioural key features and their incorporation into a housing design for pigs. *Ann. Vet. Res.* 1984, *15*, 287–299.
44. Dawkins, M.S. Battery hens name their price: Consumer demand theory and the measurement of ethological 'needs'. *Anim. Behav.* 1983, *31*, 1195–1205. [CrossRef]
45. Matthews, L.R.; Ladewig, J. Environmental requirements of pigs measured by behavioural demand functions. *Anim. Behav.* 1994, *47*, 713–719. [CrossRef]
46. Hughes, B.O.; Duncan, I.J.H. The notion of ethological 'need', models of motivation and animal welfare. *Anim. Behav.* 1988, *36*, 1696–1707. [CrossRef]

47. Lawrence, A. Consumer demand theory and the assessment of animal welfare. *Anim. Behav.* **1987**, *35*, 293–295. [CrossRef]
48. McCulloch, S.P. A Critique of FAWC's Five Freedoms as a Framework for the Analysis of Animal Welfare. *J. Agric. Environ. Ethics* **2013**, *26*, 959–975. [CrossRef]
49. Beter Leven. Available online: https://beterleven.dierenbescherming.nlhome (accessed on 5 August 2019).
50. McMillan, F.D. Quality of life in animals. *J. Am. Vet. Med. Assoc.* **2000**, *216*, 1904–1910. [CrossRef]
51. Yeates, J. Quality of life and animal behaviour. *Appl. Anim. Behav. Sci.* **2016**, *181*, 19–26. [CrossRef]
52. Skevington, S.M.; Böhnke, J.R. How is subjective well-being related to quality of life? Do we need two concepts and both measures? *Soc. Sci. Med.* **2018**, *206*, 22–30. [CrossRef] [PubMed]
53. Vøls, K.K.; Heden, M.A.; Kristensen, A.T.; Sandøe, P. Quality of life assessment in dogs and cats receiving chemotherapy—A review of current methods. *Vet. Comp. Oncol.* **2017**, *15*, 684–691. [CrossRef] [PubMed]
54. Silkstone, M.; Brudzynski, S.M. The antagonistic relationship between aversive and appetitive emotional states in rats as studied by pharmacologically-induced ultrasonic vocalization from the nucleus accumbens and lateral septum. *Pharmacol. Biochem. Behav.* **2019**, *181*, 77–85. [CrossRef] [PubMed]
55. Seligman, M.E.P.; Steen, T.A.; Park, N.; Peterson, C. Positive psychology progress: Empirical validation of interventions. *Am. Psychol.* **2005**, *60*, 410–421. [CrossRef] [PubMed]
56. Porter, D.G. Ethical scores for animal experiments. *Nature* **1992**, *356*, 101–102. [CrossRef] [PubMed]
57. Stafleu, F.R.; Tramper, R.; Vorstenbosch, J.; Joles, J.A. The ethical acceptability of animal experiments: A proposal for a system to support decision-making. *Lab. Anim.* **1999**, *33*, 295–303. [CrossRef] [PubMed]
58. Capdeville, J.; Veissier, I. A Method of Assessing Welfare in Loose Housed Dairy Cows at Farm Level, Focusing on Animal Observations. *Acta Agric. Scand. Sect. A Anim. Sci.* **2001**, *51*, 62–68. [CrossRef]
59. Spoolder, H.; Rosa, G.D.; Hörning, B.; Waiblinger, S.; Wemelsfelder, F. Integrating parameters to assess on-farm welfare. *Anim. Welf.* **2003**, *12*, 529–534.
60. Keeling, L.J. *An Overview of the Development of the Welfare Quality Project Assessment Systems*; Cardiff University: Cardiff, UK, 2009; pp. 1–110.
61. Sandøe, P.; Corr, S.; Lund, T.; Forkman, B. Aggregating animal welfare indicators: Can it be done in a transparent and ethically robust way? *Anim. Welf.* **2019**, *28*, 67–76. [CrossRef]
62. Sandøe, P.; Forkman, B.; Hakansson, F.; Andreasen, S.N.; Nøhr, R.; Denwood, M.; Lund, T.B. Should the Contribution of One Additional Lame Cow Depend on How Many Other Cows on the Farm Are Lame? *Animals* **2017**, *7*, 96. [CrossRef]
63. De Graaf, S.; Ampe, B.; Winckler, C.; Radeski, M.; Mounier, L.; Kirchner, M.K.; Haskell, M.J.; Van Eerdenburg, F.J.C.M.; Des Roches, A.D.B.; Andreasen, S.N.; et al. Trained-user opinion about Welfare Quality measures and integrated scoring of dairy cattle welfare. *J. Dairy Sci.* **2017**, *100*, 6376–6388. [CrossRef] [PubMed]
64. Tuyttens, F.A.M.; Vanhonacker, F.; Van Poucke, E.; Verbeke, W. Quantitative verification of the correspondence between the Welfare Quality®operational definition of farm animal welfare and the opinion of Flemish farmers, citizens and vegetarians. *Livest. Sci.* **2010**, *131*, 108–114. [CrossRef]
65. Diener, E.; Suh, E.M.; Lucas, R.E.; Smith, H.L. Subjective well-being: Three decades of progress. *Psychol. Bull.* **1999**, *125*, 276–302. [CrossRef]
66. King, J.E.; Landau, V.I. Can chimpanzee (Pan troglodytes) happiness be estimated by human raters? *J. Res. Personal.* **2003**, *37*, 1–15. [CrossRef]
67. Adams, M.J.; King, J.E.; Weiss, A. The majority of genetic variation in orangutan personality and subjective well-being is nonadditive. *Behav. Genet.* **2012**, *42*, 675–686. [CrossRef]
68. Weiss, A.; King, J.E.; Inoue-Murayama, M.; Matsuzawa, T.; Oswald, A.J. Evidence for a midlife crisis in great apes consistent with the U-shape in human well-being. *Proc. Natl. Acad. Sci. USA* **2012**, *109*, 19871–19872. [CrossRef] [PubMed]
69. Weiss, A.; Adams, M.J.; King, J.E. Happy orang-utans live longer lives. *Biol. Lett.* **2011**, *7*, 872–874. [CrossRef]
70. Robinson, L.M.; Altschul, D.M.; Wallace, E.K.; Úbeda, Y.; Llorente, M.; Machanda, Z.; Slocombe, K.E.; Leach, M.C.; Waran, N.K.; Weiss, A. Chimpanzees with positive welfare are happier, extraverted, and emotionally stable. *Appl. Anim. Behav. Sci.* **2017**, *191*, 90–97. [CrossRef]
71. Bethell, E.J.; Koyama, N.F. Happy hamsters? Enrichment induces positive judgement bias for mildly (but not truly) ambiguous cues to reward and punishment in Mesocricetus auratus. *R. Soc. Open Sci.* **2015**, *2*, 140399. [CrossRef]

72. Webb, L.E.; Veenhoven, R.; Harfeld, J.L.; Jensen, M.B. What is animal happiness? *Ann. N. Y. Acad. Sci.* **2018**, *1438*, 62–76. [CrossRef]
73. Mellor, D.; Beausoleil, N. Extending the "Five Domains" model for animal welfare assessment to incorporate positive welfare states. *Anim. Welf.* **2015**, *24*, 241–253. [CrossRef]
74. Boissy, A.; Erhard, H.W. How Studying Interactions Between Animal Emotions, Cognition, and Personality Can Contribute to Improve Farm Animal Welfare. In *Genetics and the Behavior of Domestic Animals*; Elsevier: London, UK, 2014; pp. 81–113.
75. Krebs, B.L.; Marrin, D.; Phelps, A.; Krol, L.; Watters, J.V. Managing Aged Animals in Zoos to Promote Positive Welfare: A Review and Future Directions. *Animals* **2018**, *8*, 116. [CrossRef] [PubMed]
76. Vigors, B. Citizens' and Farmers' Framing of 'Positive Animal Welfare' and the Implications for Framing Positive Welfare in Communication. *Animals* **2019**, *9*, 147. [CrossRef] [PubMed]

© 2019 by the authors. Licensee MDPI, Basel, Switzerland. This article is an open access article distributed under the terms and conditions of the Creative Commons Attribution (CC BY) license (http://creativecommons.org/licenses/by/4.0/).

Review

How Can We Assess Positive Welfare in Ruminants?

Silvana Mattiello [1,*], Monica Battini [1], Giuseppe De Rosa [2], Fabio Napolitano [3] and Cathy Dwyer [4]

[1] Dipartimento di Scienze Agrarie e Ambientali - Produzione, Territorio, Agroenergia, Università degli Studi di Milano, 20133 Milan, Italy; monica.battini@unimi.it
[2] Dipartimento di Agraria, Università degli Studi di Napoli Federico II, 80055 Portici, Italy; giuseppe.derosa@unina.it
[3] Scuola di Scienze Agrarie, Forestali, Alimentari ed Ambientali, Università degli Studi della Basilicata, 85100 Potenza, Italy; fabio.napolitano@unibas.it
[4] Animal Behaviour and Welfare, SRUC, Easter Bush Campus, Edinburgh EH9 3JG, UK; Cathy.Dwyer@sruc.ac.uk
* Correspondence: silvana.mattiello@unimi.it

Received: 26 July 2019; Accepted: 30 September 2019; Published: 2 October 2019

Simple Summary: The concern for better farm animal welfare has been greatly increasing among scientists, veterinarians, farmers, consumers, and the general public over many years. As a consequence, several indicators have been developed to assess animal welfare, and several specific protocols have been proposed for welfare evaluation. Most of the indicators developed so far focus on the negative aspects of animal welfare (e.g., lameness, lesions, diseases, presence of abnormal behaviours, high levels of stress hormones, and many more). However, the lack of negative welfare conditions does not necessarily mean that animals are in good welfare and have a good quality of life. To guarantee high welfare standards, animals should experience positive conditions that allow them to live a life that is really worth living. We reviewed the existing indicators of positive welfare for farmed ruminants and identified some gaps that still require work, especially in the domains of Nutrition and Health, and the need for further refinement of some of the existing indicators.

Abstract: Until now, most research has focused on the development of indicators of negative welfare, and relatively few studies provide information on valid, reliable, and feasible indicators addressing positive aspects of animal welfare. However, a lack of suffering does not guarantee that animals are experiencing a positive welfare state. The aim of the present review is to identify promising valid and reliable animal-based indicators for the assessment of positive welfare that might be included in welfare assessment protocols for ruminants, and to discuss them in the light of the five domains model, highlighting possible gaps to be filled by future research. Based on the existing literature in the main databases, each indicator was evaluated in terms of its validity, reliability, and on-farm feasibility. Some valid indicators were identified, but a lot of the validity evidence is based on their absence when a negative situation is present; furthermore, only a few indicators are available in the domains of Nutrition and Health. Reliability has been seldom addressed. On-farm feasibility could be increased by developing specific sampling strategies and/or relying on the use of video- or automatic-recording devices. In conclusion, several indicators are potentially available (e.g., synchronisation of lying and feeding, coat or fleece condition, qualitative behaviour assessment), but further research is required.

Keywords: ruminants; cattle; sheep; goats; buffaloes; animal welfare; positive indicators; five domains

1. Introduction

Animal welfare research has led to a better understanding of animal welfare needs and the development of scientific welfare indicators, which have been merged into welfare assessment protocols for various species, including cattle [1], goats [2,3], and sheep [4,5]. These protocols, developed in Europe for the evaluation of animal welfare on farms, include a selection of valid and reliable indicators whose on-farm evaluation is feasible and, whenever possible, follows European Food Safety Authority's recommendations [6] that indicators are animal-based, whereas resource- and management–based indicators are considered as "risk factors".

On-farm welfare assessment schemes focus almost exclusively on the evaluation of negative welfare indicators (e.g., lameness, overgrown claws, lesions, abnormal behaviour, excessive aggressiveness, fear) and provide an output on welfare levels based on the quantification of such negative aspects: if the presence of negative indicators is frequent, the level of welfare is low, and vice versa. In this view, the lack of suffering and the satisfaction of animals' basic requirements is indicative of a good welfare level. This follows the earlier concepts of animal welfare, based on the respect of the Five Freedoms deriving from the Brambell Report [7], formalised by the Farm Animal Welfare Council (FAWC) in 1979, and later revised and translated into the Four Principles and Twelve Criteria during the Welfare Quality® EC Project [1]. These concepts are the drivers of most of the current legislation and codes of practice on animal welfare, essentially based on the avoidance of unnecessary suffering.

However, a lack of suffering does not guarantee that animals are experiencing a positive welfare state. Several studies have argued for the inclusion of positive welfare indicators or consideration of the positive aspects of animal welfare as well as the negative in achieving a more comprehensive view of animal welfare [8]. Animals are motivated to gain a resource or achieve a particular interaction, and the affective state of achieving these goals or the reward is pleasure. Thus, what animals want, in terms of seeking positive or attractive stimuli, are associated with animals "wanting" these rewards [8]. The FAWC [9] suggests that the minimum welfare level should be defined in terms of an animal's Quality of Life (QoL) over its whole lifetime, and that QoL can range from a very poor level (a life not worth living), to a medium level (a life worth living), up to a high level (a good life).

Given that animal welfare is not only an absence of negative states (e.g., pain, disease, fear), and something positive should be provided to farm animals to make their lives worth living, positive animal welfare remains difficult to define. Previous constructs of animal welfare suggest that welfare can be defined by concerns falling into one of three domains: biological functioning, naturalness, and the feelings of affective experiences of the animal [10]. Following the functional approach, positive welfare may just be something that goes beyond the provision of farming conditions allowing the animals to be in good health. Grazing may be an example: ruminants can survive and be healthy without expressing this behaviour; hence, access to pasture may be considered a benefit and an indicator of positive welfare [11]. However, according to the natural approach, ruminants should be free to express any species-specific behaviour; thus no grazing would be a sign of a negative state, whereas access to pasture would be just the normal condition. The focus could then be on animal feelings, although short-term preferences may not match animal long-term interests. However, according to [12], good welfare can be achieved when animals are able to have a certain degree of control over the surrounding environment while tackling and meeting the challenges. In other words, a good life is not a life without challenges with both too high and too low levels of stimulation possibly perceived as aversive. Animals have expectations about the environment where they live and [13] stated that a positive experience occurs when an animal "actively responds to motivations to engage in rewarding behaviours, including all associated appetitive and consummatory effects that are positive". For example, if an animal is 'fully engaged in exploring and food gathering in a stimulus-rich environment and interacts pleasantly with other animals in the social group', then that animal may be considered to be in a positive welfare state. Both views highlight the active role played by the animals while positively interacting with the surrounding environment.

Therefore, good animal welfare should also be considered in the light of the presence of positive experiences or sensations, and not only as the result of the absence of negative experiences [14], and the balance between positive and negative effects should be in favour of the former [15]. This implies moving from the concept of "Freedoms" towards the concept of "Provisions", as animals should be managed in order to provide them with a range of opportunities to experience comfort, pleasure, interest, confidence, and a sense of control [15].

In line with these considerations, the OIE Terrestrial Animal Health Code [16] recently stated that "Animal welfare means how an animal is coping with the conditions in which it lives. An animal is in a good state of welfare if (as indicated by scientific evidence) it is healthy, comfortable, well-nourished, safe, able to express innate behaviour, and if it is not suffering from unpleasant states such as pain, fear, and distress". This definition clearly emphasises the need for positive experiences (health, comfort, good nutrition, and freedom to express natural behaviour) in the first place and mentions the freedom from suffering only at the end.

However, most research has focused on the development of indicators of negative welfare, and relatively few studies provide information on valid, reliable, and feasible indicators addressing positive aspects of animal welfare. The identification of animal-based indicators with these characteristics would allow their inclusion in welfare assessment schemes. This would be beneficial not only for improving the level of animal welfare, leading to a high QoL of farmed animals but also for reinforcing the communication about animal welfare to the stakeholders, who are strongly interested in positive indicators [17].

Mellor [17–19] proposes a five domains model to draw attention to important areas deserving consideration when talking about animal welfare. The model takes into consideration four domains related to internal states and external circumstances (i.e., Nutrition, Environment, Health, and Behaviour), and a fifth domain, i.e., Mental State, which is a final component showing positive or negative affective engagement resulting from the sum of internal states and external circumstances.

The aim of the present review is to identify existing valid and reliable animal-based indicators for the assessment of positive welfare that might be included in welfare assessment protocols for ruminants, and to discuss them in the light of the five domains model, highlighting possible gaps to be filled by future research.

2. Materials and Methods

As a starting point, an extensive review of scientific literature was carried out in the main databases (Web of Science, CAB Abstracts, PubMed, and Scopus), using keywords such as "positive welfare", "measure", "indicator", "comfort", "human-animal relationship", "emotions", "natural behaviour", "pleasure", "liveliness", "synchronization", "play" combined with "ruminant", "cattle", "cow", "sheep", "goat", and "buffalo". A total of 45 records, including 12 reviews, were obtained from this initial search. On the basis of the references cited in these records, and of the suggestions from the reviewers of the initial version of this manuscript, we enlarged our search, to obtain the final list included in this review. Only English language studies published in international journals, international book chapters or international protocols were retained. We focused exclusively on animal-based indicators that could be collected on-farm. Animal-based indicators requiring subsequent laboratory analysis were discarded. Publications dealing with resource- and management-based indicators were also excluded.

Based on the existing literature, each indicator was evaluated in terms of its validity, reliability (test–retest reliability, intra- and inter-observer reliability), and on-farm feasibility (Table 1).

Resource- and management-based measures were excluded, and only animal-based measures were considered, as this approach seems more appropriate for measuring the actual welfare state of the animals and the way in which they respond to the farming environment [6].

The results are presented in the light of the five domains considering: Nutrition, Environment, Health, Behaviour, and the overall affective Mental State [17].

Some indicators can provide useful information related to more than one domain; when this was the case, it was specifically mentioned in relation to each domain.

Table 1. Definitions of terms proposed by Battini et al. [20] and used in the present review to describe the characteristics of the considered indicators.

Term	Definition
Validity	The relation between a variable and what it is supposed to measure or predict. It can be shown by the ability of an indicator to predict some later criterion, such as a state of pleasure, comfort, vitality, etc. (predictive validity), or by the correlation between an indicator and other measures to which it is theoretically related (i.e., gold standard) (concurrent validity)
Reliability	The extent to which a measurement is repeatable and consistent
Test–retest reliability	The extent to which a measurement is repeatable and consistent throughout time
Intra-observer reliability	The agreement between successive observations of the same individual or group by a single observer, based on statistical significance of correlations ($p < 0.05$) or to Kendall's coefficient of concordance (>0.7). According to time between measurements, reliability may be classified in short- (1–7 days), medium- (1 week to 1 month), or long-term reliability (>1 month)
Inter-observer reliability	The agreement between different observers during a simultaneous observation, based on statistical significance of correlations ($p < 0.05$) or to Kendall's coefficient of concordance (>0.7)
On-farm feasibility	The practical chance of using the indicators during on-farm inspection. It may consider different constraints, e.g., time, cost, accessibility, equipment requirements, no laboratory analysis

3. Promising Indicators in the Five Domains

A list of potential indicators of positive welfare indicators in Mellor's four domains [17–19] related to internal states and external circumstances (i.e., Nutrition, Environment, Health, and Behaviour) is summarised in Table 2.

Table 2. List of the reviewed potential positive welfare indicators related to internal states and external circumstances (i.e., Nutrition, Environment, Health, and Behaviour). The animal category and method used for data collection are also specified for each indicator.

Provisions	Welfare Indicator	Animal Category	Data Collection Method	References
	Expression of feeding preferences	Sheep	Direct observations	[21]
	Grazing behaviour	Beef cattle	Direct observations	[22]
	Synchronisation of feeding	Beef cattle	Direct observations	[23]
	Bipedal stance	Goat kid	Direct observations	[24]
	Climbing	Goat	Video recording	[25]
	Comfort index	Dairy cow	Video recording	[26]
	Duration of lying bouts	Dairy cow	Video recording	[27]
		Dairy cow	Electronic device	[28]
		Calves	Video recording	[29]
	Duration of lying time	Dairy cow	Video recording	[27]
		Dairy cow	Electronic device	[28]
		Dairy cow	Video recording	[30]
		Dairy cow	Video recording	[31]
		Goat	Video recording	[32]
Environment		Heifer	Direct observations	[33]
		Sheep	Video recording	[34]
		Sheep	Video recording	[35]
	Exploration/chewing of branches	Goat	Direct observations	[36]
	Frequency of lying bouts	Dairy cow	Video recording	[27]
		Dairy cow	Electronic device	[28]
		Heifer	Direct observations	[33]
	Licking while standing on 3 legs	Dairy cow	Video recording	[37]
	Lying posture (sternal recumbency with head against the flank, in lateral recumbency with stretched legs, lying fully stretched)	Dairy cow	Direct observations	[38]
		Dairy cow	Direct observations	[39]
		Dairy cow	Direct observations	[40]
	Nibbling on objects	Goat	Video recording	[25]

127

Table 2. Cont.

Provisions	Welfare Indicator	Animal Category	Data Collection Method	References
	Playing	Goat	Direct observations	[36]
		Goat kid	Direct observations	[24]
	Ruminating while lying	Dairy cow	Direct observations	[40]
	Step up on an object	Goat kid	Direct observations	[24]
		Dairy cow	Direct observations	[41]
	Synchronisation of lying	Goat	Video recording	[32]
		Heifer	Direct observations	[33]
		Sheep	Video recording	[34]
	Time lying by a wall	Goat	Video recording	[32]
		Sheep	Video recording	[33]
	Use of brush	Goat	Video recording	[25]
	Fleece quality	Sheep		[3]
Health	Hair coat condition	Dairy goats	Direct observations	[42]
	Months staying in the herd	Dairy cow	Direct observations	[43]
	Vigour score	Lambs	Direct observations	[44,45]
	Allogrooming	Dairy cow	Video recording	[46]
		Dairy cow	Video recording	[47]
		Beef cattle	Direct observations	[48]
	Avoidance distance at feeding place	Buffalo	Video recording	[49]
Behaviour		Dairy cow	Direct observations	[50]
	Avoidance distance in the barn	Dairy cow	Direct observations	[51]
	Exploration	Beef cattle	Direct observations	[52]
	Licking while standing on 3 legs	Dairy cow	Video recording	[37]
	Locomotor play	Veal calf	Video recording	[53]
	Percentage of animals in the mud	Buffalo	Direct observations	[54]
	Self-grooming	Beef cattle	Direct observations	[55]
		Veal calf	Video recording	[56]
	Synchronisation of behaviours	Dairy cow	Direct observations	[41]
		Beef cattle	Direct observations	[23]

3.1. Nutrition

Positive aspects of welfare associated with the nutrition domain go beyond the bare satisfaction of physiological nutritional requirements, and imply, for example, aspects of choice and variety of food with pleasant smell, taste, and texture, pleasures associated with active engagement and exploration of the environment during foraging, oral pleasures of chewing/sucking, or hedonic properties of food, that eventually lead to a positive mental state [18]. The positive welfare aspects of nutrition would include measures indicative of the hedonic pleasures associated with consuming preferred foods, sensory aspects of pleasurable tastes [57] and the quenching aspects of drinking, as well as the feelings of satiation and the anticipatory pleasures of seeking and consuming food. According to [58], the possibility of food choice provides animals with the freedom to express their normal behaviour, to meet specific individual needs, and also to reduce the incidence of illness by better coping with toxins and parasite loads. Ruminants express feeding preferences if they are allowed to select their diet without any constraints, whereas the diet they actually select represents their preferences as modified by any environmental factors (e.g., accessibility, competition with conspecifics) [59]. Ruminants show feeding preferences based on forage taste, odour, and texture characteristics because these animals can associate such characteristics with the post-ingestion effects of feeds at the gastro-intestinal level [60]. In particular, they are able to avoid unpalatable feeds with low nutrient content or high toxin levels while actively selecting palatable and nutritious feeds with the aim to maximize their nutritional well-being [61]. Catanese et al. [62] confirmed that limiting diet choice induces a stress response in lambs. At least in humans, a close relationship has been found between affect and food consumption (e.g., [63]). Therefore, it has been postulated that also in non-human animals, the ingestion of pleasant feeds, based on their hedonic values, may induce a more positive affective state as compared to receiving less palatable feeds. In fact, [21] observed that after the consumption of a pleasant pellet, ewes show an optimistic judgment bias by approaching non-reinforced ambiguous locations more quickly (i.e., the ewes received no training about the possible presence of palatable or unpalatable feeds in those locations) than ewes receiving disliked wood pellets. These results indicate the expression of feeding preferences as a potentially valid indicator, although no evidence is available on its reliability, and preference tests may not be easily performed on-farm.

In rodents, behavioural indicators of the pleasure of eating or drinking have been quantified as tongue protrusions, lip smacking, and lick patterns (e.g., [64]), but similar animal-based indicators have not been developed for ruminants. Both sheep and cattle consistently show a preference of legumes over grasses (1.5 times higher intake) and also a particular diurnal pattern, with legumes preferred in the morning and preference for grasses increasing at the end of the day [59]. These data suggest that ruminants have specific goals when selecting their diet, which cannot be accomplished when they are fed a total mixed ration, with no possible alternative. However, it should be considered that feed choice may be driven by individual differences [65], as some animals are more prone to consume a regular and constant diet, whereas others are willing to explore new feeds, as recorded in heifers by Meagher et al. [66]. It may be argued that, if animals are given the possibility of choosing their diets, they may not necessarily eat what is best for them, and they may not consume a diet adequate to meet their nutritional requirements. Although we cannot exclude that this may happen in some cases, research by [67] showed that calves offered with a varied diet reached the same nutritional level provided by a standard balanced mixed ration, yet each animal ate a diet different from the other animals. Data supporting the evidence that ruminants are able to select a diet close to their needs and minimise the ingestion of anti-nutritional compounds when they are given the possibility to choose among different feeds are reviewed by [58].

An additional benefit in terms of welfare can be given to ruminant animals by allowing the expression of their dietary preferences while performing their species-specific grazing behaviours (i.e., exploration, selection, and ingestion of plants [22]), as also envisaged by the Welfare Quality protocol [1,12].

Domestic ruminants are gregarious animals and their feeding behaviour, as well as other maintenance behaviours [68–70], is synchronised [71]. Feeding behaviour synchronisation in social animals is an adaptive behaviour, evolved to provide a series of benefits, such as opportunities for acquiring information about the location of food and allowing more time to graze, due to reduced exposure to predation risk [72]. In this sense, feeding synchronisation may indicate a positive welfare condition for all group members (see also "Behaviour"), and it was actually observed more frequently in finishing bulls kept on pasture than in a more restricted housing environment [23]. Furthermore, feeding synchronisation may indicate reduced competitiveness, thus representing an additional benefit for subordinate animals, as they would be able to access the feeding resources along with conspecifics. The assessment of the synchrony of feeding is feasible on-farm and it could be achieved by instantaneous scan sampling [68], taking into account that synchronisation is maximal in the morning and in the evening [73]. No information is available about the reliability of feeding synchronisation, but behavioural synchronisation is generally considered reliable [39], whereas the articles mentioned previously [23,39,70–73] support the validity of the measure.

3.2. Environment

The environment can have marked effects on animal welfare, and some literature suggests the positive effects of housing enrichments, which should be beyond the simple housing supplementation (i.e., just capable of reducing the negative effects of a poor environment). Examples of housing enrichment are reviewed by [74–76]. Positive aspects of housing or the environment involve providing the animal with the space and requirements for comfort and pleasure associated with resting and ease of movement, as well as offering choice and opportunity to express agency in use of the environment.

Comfort and appropriate rest are important components of positive welfare induced by the environment. For example, reduced lying time and abnormal lying postures or transition movements are shown when housing conditions are suboptimal (e.g., [77,78]). We can, therefore, argue that an increase of lying time, the possibility to perform appropriate lying postures, and the ease to get up and lie down may be indicative of a positive welfare state.

In ruminants (e.g.,: cattle: [34,79]; sheep: [34]), resting is a high priority and an inelastic behavioural need. The total amount of time spent lying was used to assess cow comfort in response to the type of housing (large pens vs tie-stalls) and the depth and shape of sand bedding, respectively [27,28]. In both studies, a higher lying time occurred in the more favourable conditions (large pens and deeper bedding), thus confirming the validity of this positive indicator to assess cow comfort during resting. The time spent lying was also observed to increase in cows, sheep, and in dairy calves provided with a more insulating substrate (e.g., sand bed vs concrete floor in cows [30]; straw bed rather than concrete or slatted flooring in shorn ewes [35]; sawdust rather than river stones in calves [29]), demonstrating lying time to be a good positive indicator for the evaluation of bedding quality and of thermal comfort. Another example that supports the validity of lying time for the evaluation of cows' comfort in relation to the environment is provided by [31], who observed an increase of the time spent lying on more comfortable lying substrates (i.e., rubber mats vs concrete or sand).

In goats, the total time spent lying was positively affected by increased indoor space allowance [80], but did not statistically vary in response to the inclusion in the pens of additional walls that had been introduced in order to increase goats' comfort and to provide a higher sense of protection not only from virtual predators, but also from higher-ranking goats. However, goats spent significantly more time resting in the resting area with wall support, suggesting that this indicator (resting by a wall) may be used as an indicator of comfort in this species [32]. In an analogous experiment in sheep, similar results were obtained on lying time, but not on time spent resting by a wall [34]. However, lying time may be increased when animals are unwell (sickness behaviour) or when lame (e.g., [81]); thus, alone, this indicator is not specific for positive welfare.

In cattle, [27] used the frequency of lying bouts as an indicator of the ease of transition movement, and validated this indicator by showing that the number of transition movements was significantly higher in large pens than in tie-stalls. As the mean duration of single lying bouts did not differ between treatments, the higher number of lying bouts resulted in a longer total lying time, which was also considered as a positive indicator of animal welfare. However, [28] could not confirm the validity of the number of lying bouts to assess cows' comfort, whereas they found significant differences in the duration of lying bouts in relation to bedding depth and shape. In contrast, [33] did not record any significant difference either in total lying time or in the number of lying bouts, in response to different space allowances.

None of the studies mentioned have investigated these indicators in terms of reliability (test-retest, inter-observer, or intra-observer reliability) and the on-farm feasibility of these indicators still has to be discussed. In fact, all the studies were based on either direct- [33] or video-recorded [27,32,34] observations lasting 24 h, which is impractical for on-farm assessment. Alternatively, electronic devices were used in the study by [28], but their use for a practical on-farm assessment is also questionable. Sampling observation rules would be required in order to increase the feasibility of these indicators. However, circadian rhythms of resting periods may show a pronounced variation, especially in extensively managed animals [81], and this may have a marked impact on the selection of sampling periods for the assessment of lying time.

Napolitano et al. [68] suggested that lying postures can also be used to highlight the level of thermal comfort and/or vigilance comfort, and state that cows prefer to rest in sternal recumbency with the head tucked against the flank, and in lateral recumbency, possibly with outstretched legs. A positive correlation was found in dairy cows between cubicle features (i.e., stall width, stall length, amount of straw, area, and type of divider) and some lying postures: stretched forelegs, stretched hindlegs, lying fully stretched [38]. van Erp-van der Kooij et al. [40] confirmed the preference of dairy cows for lying in long and wide postures when at pasture, in a comfortable situation. These authors also carried out a reliability analysis on lying postures at the start of their study, finding an initial moderate agreement (kappa values 0.49 and 0.50) between a trainer and the observers, whereas [39] found good inter-observer reliability and consistency over time for head resting and hind legs stretched. Also, for the observation of lying postures, the development of a representative sampling strategy may be required for practical on-farm observations.

Another important aspect of lying behaviour is the level of synchronisation (see also "Behaviour"), that may be considered as indicative of a positive welfare state. For example, Holstein heifers with larger space allowance exhibited a higher synchronisation of lying behaviour, which was interpreted as a higher welfare level, thus confirming the predictive validity of this indicator [33]. The validity of the synchronisation of resting behaviour could not be confirmed as an indicator of comfort in sheep [34] and goats [32], which did not increase their synchrony after the inclusion of additional walls. However, the presence of additional walls is only one of the factors that can potentially affect lying behaviour: the space available in these studies (1.5 m^2/head, which corresponds to the minimum recommended value for small ruminants; [69]) may have contributed to the inability of all animals lying down at the same time, independently from the presence of partitions. In fact, [81] observed that sheep are more synchronised when they have more space to lie in an indoor environment. Based on the concept of synchronised lying, [26] proposed the use of a Comfort index, calculated as the number of cows lying in free stalls out of the total number of cows touching a stall. According to these authors, a value greater than 85% is considered a good threshold. The validity of lying synchrony as a positive welfare indicator is confirmed by [38], who found a positive correlation between the maximum synchronous lying and cubicle features (i.e., stall width, stall length, amount of straw, area, and type of divider) providing more space and a more comfortable and insulated lying surface. The same author also found a positive correlation of these features with the percentage of cows ruminating while lying: this indicator can also be considered as indicative of a positive welfare state, as rumination is usually performed by healthy, relaxed, and unstressed cows while lying down [82]. The inter-observer reliability and consistency over time of lying synchrony have been confirmed by [39] both for direct and video-recorded observations.

As for feeding synchrony, also lying synchrony can be assessed using instantaneous scan sampling: this indicator can be collected in a quicker way than lying time, and it is less likely affected by circadian behavioural changes [81].

Another behaviour that may indicate that animals are living in a good environment is exploration, which usually increases in novel and complex environments, as discussed in the paragraph "Behaviour" (e.g., [54]).

3.3. Health

Traditionally, assessment of health aspects of welfare has focused on categorising and auditing the most common health issues of the species (number of lame animals, parasitised, showing visible injuries, and so on). However, the absence of clinical disease or injury is not the same as the positive aspects of good health, such as feeling well, active, and vigorous. Suitable positive welfare indicators in this domain would include measures that suggest animals are enjoying vitality and good health.

The longevity of breeding animals can be defined as the time-span animals remain in the herd, whereas productive longevity is the period between first parturition and culling. In intensively farmed dairy cow herds, longevity and productive longevity are usually well below five and three years, respectively [43], whereas cows kept in extensive conditions or in small family farms show a mean longevity of 15 years [83] and cases are reported where animals reach over 22 years [84]. Most of the factors that lead to the culling of dairy cows concern health (e.g., mastitis, lameness, low fertility) and unsatisfactory production. In farm animals, a long life may be considered as the result of good welfare [85], and longevity as its "summary indicator" (i.e., summarising all the potential noxious factors leading to a reduced life expectancy). However, it may be speculated that the validity of this indicator relies on the reliability of the information about culling reasons, as only those concerning involuntary culling should be considered. In addition, there may also be concerns about Quality of Life, as just being on the farm for a long time may not mean positive welfare: a longer time spent in pain but not reaching the point of needing to be culled may have obvious negative effects on the welfare of the animals, and may result in a life that is not "a good life", and possibly not even "a life worth living" [85].

Vigour is another indicator of positive health, which expresses positive and active engagement with the environment. In lambs, a vigour score has been proposed, based on the latency to first perform specific behaviours, such as an attempt to stand, seeking the udder, and successful sucking [44,45]. This score has been shown to be valid, in that it is reflective of the behaviour of neonatal lambs [45], and has been applied on commercial farms [45] suggesting data can be collected feasibly, at least on farms with indoor lambing systems. The reliability of scoring has not been formally tested.

A positive health condition can also be identified by coat or fleece conditions. For example, the percentage of goats with coat described as "a complete fur cover, even coat, presenting shiny, glossy and sheen hair, homogeneous and well adherent to the body" [42] has been included in the AWIN welfare assessment protocol for goats [3], and a good fleece quality (sufficient fleece, no trailing or over long patches of fleece, no scurf nor lumpiness, nor evidence of ectoparasites) was included in the AWIN welfare assessment protocol for sheep [5]. For hair coat condition in goats, [42] demonstrated high inter-observer reliability, but [86] showed a low consistency over time. For sheep, assessments of fleece quality have good intra-observer, but poor inter-observer reliability, when assessed at a group level [87], but good inter-observer reliability when assessed individually (AWIN 2015, unpublished).

3.4. Behaviour

Positive welfare may be evident when animals are able to express active and positive engagement with the environment and in their interactions with other animals, resulting in exploration, foraging, hunting, bonding, affiliative social contacts (such as play, social grooming, and other pleasurable contacts,) and positive parent-offspring interactions [13,88].

Numerous authors report that human and non-human mammals play when they are not exposed to harmful events and threats to fitness, such as abrupt weaning, insufficient nutrient intake [89], disbudding [53], and castration [90]. This may be explained by the fact that this behaviour is not needed for survival; thus, it is not expressed in unfavourable conditions. According to [91], play is self-rewarding, may be included in the behavioural repertoire of the adults albeit at a lower frequency than juveniles of the same species, and it is repetitive, although not stereotypical in form. In addition, play has been associated with positive emotions [92]. Valnickova et al. [93] observed that deprivation of play induced reduced growth in dairy calves, thus suggesting the validity of the absence of this behaviour as an indicator of negative welfare. However, a rebound effect in play is often observed when conditions improve. For example, the play expressed by dairy cows when, during winter housing, they are released into exercise areas, may not be an indication of positive welfare, rather, it may represent a sign of relief from a previous poorer condition [94]. In general, it can be stated that reductions in play are associated with negative affect, whereas there is evidence for an increase in play with positive affect. Anticipatory behaviours can provide information about emotional states and the anticipation of play, being a rewarding event, can be a positive state as shown by an increased frequency of behavioural transitions and duration of walking [95]. However, inferences on anticipatory behaviours should be considered, as long waiting periods may induce frustration [96]. The registration of spontaneous play in young ruminants may not be feasible due to the low level of expression [93]. Three main categories of play have been described: social, locomotor, and object play [97]. Mintline et al. [53] suggested using an arena test to elicit locomotor play and reduce the time needed to record this behavioural expression. These authors noted that the amount of time devoted to locomotor play in the home pen was positively correlated with the time spent playing in the arena test, thus suggesting the validity of the test. However, some open questions remain concerning the feasibility of including this in on-farm assessments, as well as the size and shape of the arena, both affecting the expression of this behaviour, the time elapsing between tests, with more time devoted to play at increasing elapsed times, and the high day–to–day expression variability. No studies on the reliability of this test are currently available.

In young mammals, the possibility of sucking from their dams may also be related to a positive welfare state, as this is a natural and highly motivated behaviour (in the wild, it is essential for survival) [98]. The need for sucking from a teat seems to be confirmed by the fact that, if this natural behaviour is limited or prevented, calves may redirect it towards other targets in the form of abnormal oral behaviour [99], while lambs show a number of behavioural, endocrine, and immune disturbances [100]. Conversely, when young mammals have the opportunity of sucking from their mothers, they perform this rewarding and appeasing behaviour, particularly following behavioural disturbances, suggesting that it has a rewarding and comforting component as well as nutritional. Although valid [22,68,69], to our knowledge, no information is currently available about the reliability of this indicator, and feasibility may be limited due to the time required for the observation of sucking behaviour.

A high level of synchronisation has been mentioned as an indicator of positive welfare in cattle [68], goats [69], and sheep [70]. In fact, this is an allelomimetic behaviour indicative of social cohesion [70]. According to [101], in socially stable groups, 90% of the individuals exhibit the same behaviour at the same time, whereas [73] established a 70% synchrony threshold in cattle at pasture.

The use of the level of behavioural synchronisation (standing, lying, and feeding) in cattle as a measure of positive welfare is supported by earlier research showing that dairy cows' behaviour is more synchronised on pasture than in tie stalls [41], and later confirmed by [23] in fattening bulls, where the synchronisation was higher in bulls at pasture than in bulls kept in pens in an uninsulated barn. In fact, as recently reviewed by [102], a loss of synchrony may be interpreted as an index of reduced welfare in housed vs pasture-based systems (and vice versa), probably due to the reduction of space allowance, the increased level of disturbance, and the higher competition for lying places.

Stoye et al. [73] observed that the level of synchrony was minimal in the middle of the day and peaked in the morning and in the evening, and suggest that the time of day should, therefore, be taken into account when this variable is used to measure animal welfare. This consideration can be important to set up appropriate sampling rules for behavioural observations and may contribute to increasing the on-farm feasibility of this indicator.

Allogrooming is defined as a licking or grooming behaviour performed between pairs, most commonly on the head, neck, and shoulder [71]. Allogrooming has been widely documented in adult [103]) and juvenile cattle [104], whereas in sheep and goats, it is mainly expressed by mothers to new-born animals [105]. In cattle, this behaviour tends to occur most around the arrival of fresh feed, and in longer bouts at night, with 5 min per day total time spent in allogrooming [46]. Allogrooming is not thought to be related to dominance hierarchies but is thought to be an expression of a close relationship [46,47], relevant for the formation and maintenance of social bonds. Affiliative behaviours have been proposed as indicators of positive welfare [92]. Receipt of allogrooming induces reduced heart rates and half-closed eyes [106]. In addition, it has been observed that the animals which receive more grooming have increased milk production and weight gain [107,108]. However, some issues still need to be resolved with respect to whether higher levels of social licking may be a mechanism to reduce tension [104]. In particular, allogrooming-dependent tension reduction was not seen experimentally [92,109], and higher licking was observed in tethered cattle vs loose cattle [110], perhaps related to the familiarity of neighbours or boredom, and in indoor vs outdoor cows, probably due to proximity [109]. On-farm feasibility of allogrooming seems low as it occurs for a relatively short period per animal, but at higher rates during feeding [46]. However, the feasibility on-farm has not been tested, so methods would need to be developed. Consequently, no studies on test–retest reliability and intra- and inter-observer reliability have been conducted.

Self-grooming is related to a broad behavioural category encompassing licking the coat with the tongue (generally restricted to cattle), rubbing or scratching with teeth (sheep), hind hoof, horns, or against environmental objects (trees, fencing, pen fixtures, etc.), including the use of brushes by dairy cows [111]. Self-grooming has also been documented in goats [112,113]. Studies on innate 'programmed grooming' suggest that this behaviour is influenced by age (greater self-grooming in young small ruminant animals; [114]). In wild sheep, this behaviour is related to hygiene as a means of removing dirtiness, ticks, and other ectoparasites, so it may be a motivated behaviour [115]. In general, self-grooming is deemed a comfort activity [116]. Platz et al. [37] found that comfort behaviours of individual hygiene in dairy cattle, such as licking while standing on three legs (licking herself with one leg raised from the floor surface) and caudal licking (licking of caudal parts of the body by concave flexion of the lumbar spine), are only performed on non-slippery flooring. The replacement of concrete slatted flooring by rubber mats may increase self-grooming up to 4-fold. According to [117], self-grooming tends to be performed more often when calves are motivated to express this behaviour, but the environment does not allow to exhibit it. However, whether self-grooming is positive or solely an expression of positive affect is questionable, since self-licking occurs at a higher rate in tie stalls than in loose housed cattle [110], and [118] reports that excessive self-licking can be observed in calves in response to deprivation situations.

In natural environments, cattle scratch and groom themselves on trees or other abrasive surfaces [119]. Therefore, the provision of brushes can be considered as an environmental enrichment [55] that stimulates the animals' natural behaviour, and brush usage seems to be a 'luxury' behaviour with low resilience [120], which therefore may indicate positive affect. According to [121], the voluntary use of cow brushes might be a useful indicator of positive welfare. Using a motivation test, [119] demonstrated that the motivation of cows to access a grooming substrate is as high as their motivation to access fresh food. In calves, choice tests suggest that brushing is perceived as a positive event, but heart rate variability is not affected [56]. Ninomiya [55] showed that beef cattle increase their self-grooming and scratching behaviour when their pen is enriched with a brush, and hypothesises that it is possible that the increased expression of this behaviour could be beneficial in terms of animal health, based on the lower occurrence of liver and intestinal diseases in enriched environments. In addition, brushing facilitates milk let-down and acceptance of milking in heifers, although this result may also be related to habituation to humans rather than to brushing, per se [122]. A complete validation of an indicator is still lacking as well as studies on reliability, but its observation seems feasible on-farm.

Exploration can be distinguished as specific and general, the former directed towards a specific object or event, the latter related to the collection of a broader range of information about the surrounding environment. Exploration has an adaptive value as it is information-gathering about feeding resources, and checking for potential hazards, thus making individuals more prepared for avoiding dangers and finding rapid escapes (e.g., in case of attacks by predators). Exploration seems to be self-rewarding, thus indicating a positive welfare state [52]. General exploration increases when animals are exposed to a novel environment, but it is also performed in known complex environments where this kind of exploration is performed to check for changes. In intensive conditions (i.e., no access to pasture), ruminants tend to extend their periods of inactivity while reducing their exploratory behaviours, thus suggesting a low degree of adaptation to an insufficiently stimulating environment (e.g., [54]). However, the recording of daily exploration is not feasible; thus, [52] proposed a novel object test at the feeding rack as a proxy to estimate the animals' explorative responses. These authors observed higher levels of specific exploration expressed by bulls kept in a barren environment, but they obtained weak results when this environment was enriched and concluded that the test was not promising. In an arena test, buffalo heifers kept indoors expressed higher levels of specific exploration towards a novel object as compared with animals kept on pasture [123]. The authors hypothesised that animals kept indoors due to the paucity of stimuli were more motivated to perform explorative activities. Therefore, although the arena test is reasonably feasible, its validity as a positive indicator is questionable, and data on reliability are lacking.

Ruminants can be exposed to human contacts, and this happens frequently, especially in dairy animals kept in intensive systems. The quantity and the quality of the human–animal relationship may have a prominent impact on the behaviour, welfare, and productivity of farm animals (e.g., [124]). For example, the use of negative interactions (e.g., shouting, forceful sticking, and slapping) may depress milk production [125], growth rate [126], as well as increase the fear of animals towards humans [127]. On the contrary, positive interactions (e.g., talking quietly, petting, and touching) may have beneficial effects on fertility of dairy cows [128] and growth rate of veal calves [129], provide social comfort [130], induce changes in heart rate and heart rate variability and oxytocin release [131,132], and elicit positive affective states [133]. For cattle, many measures have been proposed to assess the quality of the human–animal relationship, ranging from the observation of both stock people and animal behaviour during routine activity (milking, handling, etc.) [128] to assessing attitudes of stock people towards animals through questionnaires [51,128]. For sheep the AWIN protocol reported a human avoidance test with a familiar person at flock level [5], whereas in cattle the most used indicator has been the avoidance distance of animals to humans, defined as the distance at which an unfamiliar observer is allowed to approach an individual animal before it moves to the side or away [51]. It can be measured either at the feeding place or in the home pen [134], the first being more feasible than the latter. However, all the tests involving a human moving towards the animals may be the results of tolerance and fear, whereas, in the tests where the animals voluntarily approach unfamiliar humans, they actively elicit a contact, which can be interpreted as positive. Examples of tests where the animals approach humans are available in cattle [135] and in goats [3,136], and in responses to separation and reunion in lambs bonded to humans [131]. However, it has been postulated that these tests may be the result of conflicting motivations, such as motivation to explore and fear [135], thus affecting the validity of this indicator. Although feasible, no studies on reliability are available.

In addition to the above-mentioned behaviours that are common to farmed ruminant species, there are some species-specific behaviours whose expression can be considered as an indication of positive welfare. For example, buffaloes are the only domestic ruminants expressing the behaviour of wallowing, which consists of covering the body surface with mud. Therefore, this can be considered a species-specific natural behaviour. Although this behaviour has received little attention, when appropriate facilities are available, buffaloes lie in potholes, ponds, or pools. In particular, previous studies report that from February to July, on average, 31% of the animals were wallowing in the mud [137], whereas this increased to 48% in mid-summer (i.e., from June to August) [54]. Buffaloes have a sparse hair coat and, consequently, a reduced number of sweat glands [138]. Therefore, they use wallowing as a means to efficiently dissipate heat, as also suggested by the higher milk production of buffaloes provided with wallowing facilities [54]. Conversely, when wallowing is denied, buffaloes tend to spend more time idling [137] and lying in the slurry (possibly trying to compensate for the lack of water) and less on exploration [54]. In addition, wallowing proved to be more efficient in heat dissipation than showers [139]. However, all these findings just suggest that buffaloes may suffer if wallowing is denied, whereas to be positive, such behaviour should be self-rewarding. One aspect suggesting this effect is that, albeit less frequently, buffaloes wallow also in winter, when a thermoregulatory motivation is lacking [137]. In addition, at least in pigs, wallowing may induce relaxation [140]. This indicator may be considered both feasible and reliable if assessed as a resource (i.e., access to pools/potholes), whereas no information is available if assessed as an expression of wallowing behaviour.

Goats are known to climb and prefer elevated places [141]. Aschwanden et al. [142] found that the provision of platforms to loose-housed goats positively affects the behaviour of animals during feeding and resting and reduces aggressions. The authors found an increase in feeding and resting bouts, while the possibility to move both in horizontal and vertical space helped to minimise agonistic interactions [142]. When given the possibility, goats actively used enriched environments (e.g., niches, platforms, brushes) and were frequently observed climbing, walking, and lying in elevated places [25]. Observations of behaviours influenced by the presence of platforms or elevated places are time-consuming; hence, feasibility is presently low. No information is given about the reliability of this indicator.

One of the most important social interactions animals engage in are the contacts between mothers and offspring. In sheep and beef cattle, where offspring remain with the mother for prolonged periods, affiliative social contacts (licking, nosing, actively and preferentially seeking one another, and lying in contact) are frequent (e.g., [143,144]), and associated with elevated oxytocin in the mother (Muir & Dwyer [145]). Maternal care can be assessed by measuring grooming, mother-offspring contact and proximity, and suckling frequency. At present, these are time-consuming and impractical in the field, but with technological developments, such as proximity sensors, these may become more feasible measures in the future.

3.5. Mental State

According to Mellor [20], this domain is the result of internal states and external circumstances that may affect welfare-relevant mental experiences. However, the aim of this review is to consider indicators that can measure the positive mental experience, as generated by other factors. The indicators included in this section are not necessarily linked to a specific provision to the animals but can serve as a general overview of the affective state of ruminants. Positive emotional states of relevance to animal welfare may include, for example, calmness, relaxation, curiosity, excitement, positive engagement, and anticipation of reward or pleasurable events. Panksepp [146] postulated that affective states are functional for the fitness of the animals as positive states inform the animals about the fact that they are coping well, whereas negative states may alert subjects to potential threats to survival.

Based on previous studies of humans, [147] postulated that animals experiencing negative emotions, such as those deriving from fear, stress, and adverse environmental conditions, will interpret ambiguous stimuli in a more pessimistic way [148,149]. Conversely, albeit less studied, more optimistic judgments are given as a consequence of positive experiences such as gentle handling [150], environmental enrichment [151], and release after an aversive treatment [152]. In particular, Doyle et al. [152] noted that ten Merino sheep exhibited a positive bias when released after a period of restraint and isolation. Similar results were obtained by Sanger et al. [153]. These authors observed that shorn sheep had a more positive judgment of ambiguous stimuli after release. These studies showed a more positive response when animals were released from a negative experience, which is assumed to generate a positive affective state, although whether this is also true when there is no prior negative stimulus is yet to be tested. Although a number of physiological measures demonstrate the validity of positive judgement bias as positive welfare indicator, which directly reflects the emotional state of the animal (core affect), on-farm feasibility of this indicator seems low due to the fact that animals have to be trained and specific and articulated tests requiring a specific set-up, rather than the observation of spontaneous behaviour in an undisturbed environment, have to be put in place. Attention bias, another class of cognitive bias, has been developed in order to improve the feasibility of measuring these cognitive effects on-farm. Attention bias describes the differential allocation of attentional resources towards one stimulus compared to others [154]. Recently, [155] proposed the attention bias test where the animals do not need to be trained as the attention towards a novel, potentially noxious stimulus is recorded the first time they encounter it. These authors noted reduced attention towards a dog in more relaxed sheep (i.e., treated with diazepam). However, in cattle, this effect was not replicated [156]. No studies on intra- and inter-observer reliability have been conducted.

Behavioural laterality has also been linked to either positive or negative animals' emotions, as the two brain hemispheres control contrasting emotions: in fact, the left hemisphere is believed to control positive emotions, whereas the right hemisphere controls negative ones, and behavioural responses are contralateral to the dominating hemisphere [157]. In ruminants, behavioural lateralisation in response to different emotion-eliciting situations has been observed in sheep that showed a higher proportion of left-lateralised ears after the exposure to a non-palatable food (wooden pellet), and a lower proportion of left-lateralised ears after the exposure to a standard feed that was considered as a positive stimulus [158]. A higher proportion of right-lateralised ears can, therefore, be supposed to indicate a positive emotional state. However, the results of this experiment were not consistent with other trials carried out by the same authors, and the validity of ear lateralisation as an indicator of sheep's emotions still has to be confirmed. This indicator seems to be feasible on-farm, but no information is available on its reliability.

Some indicators may be used to interpret emotions according to a dimensional theory [159] where emotions are described as moving in a continuum along two axes: valence that expresses positive or negative moods, and arousal that defines the low or high level of excitement. Negative emotions often go along with high arousal and positive emotions with low arousal, but high arousal can sometimes also be found in positive situations. The qualitative assessment of animal behaviour (QBA) has been widely used to describe how animals interact with the environment. It is based on an integrative approach where the 'whole-animal' is assessed, according to the above-mentioned dimensional theory. This methodology relies on the use of behavioural descriptors ranging from low (e.g., calm, relaxed) to high levels of arousal (e.g., active, restless) and from a positive (e.g., curious, excited) to a negative valence/mood (e.g., indifferent, bored). These descriptors, in the original version, were generated by the observers using the free choice profile technique [160]. The approach based on free choice profile showed that QBA had high intra-observer reliability, inter-observer reliability (e.g., [160]) and validity (e.g., [161]). However, in order to make the methodology suitable for on-farm welfare assessment, fixed lists of terms have also been used and show good inter-observer reliability in cattle [162] and sheep [163]; test-retest and intra-observer reliability, as well as on-farm feasibility, have also been confirmed in sheep [164]. The validity of QBA to highlight positive emotional states has been confirmed in cattle [165], goats [166], and sheep [164].

Farm animals may also express emotions using a complex set of facial expressions, body postures, and vocalisations. Most studies focused on the assessment of these indicators in relation to pain or fear, but they are rarely adopted to assess positive emotions, such as relaxation or pleasure. Behavioural studies and the physiological basis of changes in visible eye white (or eye aperture) suggest that this indicator may reflect emotional experience in cows [167–171] and sheep [172–175]. In particular, changes in eye aperture can be a dynamic indicator of emotional states, with a low percentage of visible white indicating satisfaction and low arousal [168]. The percentage of eye white decreases when cow and calf are reunited after separation [170], sheep brushed by a familiar human show a high proportion of closed and half-closed eyes during and post brushing, indicating that this procedure might have elicited a relaxed state [172,173], and groomed sheep show low relative eye aperture [175]. However, the percentage of eye white in dairy cows may also increase in response to a positive stimulus presumed to be particularly exciting [171], including exposure to concentrate [169]. Hence, the visible eye white may indicate arousal perhaps more than valence. Validity is uncertain in several studies [169,171,174], whereas inter-observer reliability has been only tested and checked in two studies [169,171]. Reefmann et al. [175] comment on feasibility, and state that determining the eye aperture and, consequently, visible eye white, is a labour-intensive task, made of a digital calculation of the white area in comparison with the whole eye. At present, feasibility is very low for on-farm assessment; however, it could be improved by performing direct observations or developing an automatic computerised calculation. A recent work [176] checked the possibility of assessing the eye white in dairy cows as classes of eye aperture, ranging from eye white clearly visible to half-closed eye: this method appears to be promising and more feasible than any computerised calculation.

Ruminants have highly developed muscles around their ears, enabling them to independently move forward and backwards very readily to express internal states and communicate [158]. Some studies support the idea that facial expressions can communicate specific emotional states, by showing that goats and sheep avoid images of conspecific faces demonstrating fear or discomfort, and are attracted to relaxed or positive facial expressions [177]. Ear postures are highly species-depending and require specific study in order to gather accurate information [178]. Hanging ears in dairy cows and sheep are associated with positive emotional states of low arousal (e.g., stroking or grooming [131,172,173,178,179]) and related to the relaxed/calm dimension of QBA [130,178]; however, whether hanging ears would be associated with a low arousal negative state (such as boredom) has not been rigorously tested. Both backwards and forwards ears may be associated either with positive or negative situations [158,178,180,181]; hence, it is not possible to accurately relate these to underlying emotional states. Frequent ear changes are associated with negative stimuli in sheep [158], but with positive stimuli in cows [178]. Further studies are needed to clarify the meaning of ear-posture changes for each species, and to check for the reliability of this indicator.

Recent research focusing on body posture as an indicator of emotion in farm animals shows that there is considerable variation among species in the meaning of the same posture [182]. Hence, body postures cannot be generalised to different species and, furthermore, it is suggested that specific postures may have a different meaning when assessed alone or combined to a whole body posture evaluation [182]. Stretching the neck, or a horizontal neck position are associated with positive emotions in cows [167,179,182], as well as tail wagging in cattle [182] and sheep [173] and tail up in goats [181]. These postures are usually associated with ears hanging down and can be easily recorded on-farm, but no specific study is available on their reliability.

Vocalisations could be considered as a direct expression of emotion in animals [92]. Little work is available to reliably identify types of calls or acoustic parameters when ruminants are exposed to positive situations; thus, general conclusions are difficult to draw. The source–filter theory of voice production suggests that vocalisations in mammals are generated by vibrations of the vocal folds ("source") that determines the frequency (or pitch; "F0"), subsequently filtered by the vocal tract ("filter"). Hence, changes in vocal production are associated with emotion-related changes in the pharynx and glottis which alter the characteristics of the sound produced [183]. Low-frequency calls have been recorded in cows, sheep, and goats in situations that elicit positive emotions [181,183,184]. Cows and sheep produce these calls with the mouth closed or partially open [183]. It has also been found that in goats, the vocal fold vibrates at a more stable rate during positive than negative emotions, resulting in more stable F0 over time [181]. However, the use of vocalisations as an indicator to measure positive emotions has some limitations. First, it relies on animals providing enough vocalisations for assessment, and then, on the ability of human assessors to evaluate sounds. The development of automatic tools for the recognition of sound would be particularly appealing to improve the feasibility and reliability of this indicator.

A list of indicators of affective engagement resulting from the sum of internal states and external circumstances (i.e., Mental State) is presented in Table 3.

Table 3. List of the reviewed potential positive welfare indicators of affective engagement resulting from the sum of internal states and external circumstances (i.e., Mental State). The animal category and method used for data collection are also specified for each indicator.

Welfare Indicator	Animal Category	Data Collection Method	References
Asymmetric ear posture	Sheep	Video recording	[173]
Axial/plane ears	Sheep	Video recording	[175]
Body posture changes	Sheep	Video recording	[173]
Closed eyes	Sheep	Video recording	[173]
Duration in each ear posture	Sheep	Video recording	[172]
Ear-posture changes	Sheep	Video recording	[172]
Ears back down	Dairy cow	Video recording	[182]
Ears back up	Dairy cow	Video recording	[182]
Ears backwards	Dairy cow	Video recording	[178]
	Dairy cow	Photos	[176]
	Sheep	Video recording	[172]
Ears hanging	Dairy cow	Video recording	[178]
	Dairy cow	Video recording	[179]
	Dairy cow	Photos	[176]
	Lamb	Video recording	[131]
	Sheep	Video recording	[158]
Half-closed eyes	Dairy cow	Photos	[176]
	Sheep	Video recording	[172]
	Sheep	Video recording	[173]
Head orientation changes	Sheep	Video recording	[173]
Infrequent ear-changes	Sheep	Video recording	[174]
Leaning into stroker	Dairy cow	Video recording	[167]
Licking stroker	Dairy cow	Video recording	[167]
Low percentage of visible eye white	Dairy cow	Video recording	[167]
	Dairy cow	Video recording	[169]
	Dairy cow	Video recording	[170]
	Dairy cow	Video recording	[168]
	Dairy cow	Video recording	[171]
	Dairy cow	Photos	[176]
	Sheep	Video recording	[174]
Low relative eye aperture	Sheep	Video recording	[175]
	Dairy cow	Photos	[176]

Table 3. *Cont.*

Welfare Indicator	Animal Category	Data Collection Method	References
Low-frequency calls	Dairy cow	Electronic device	[183]
	Goat	Video recording	[181]
Neck horizontal	Dairy cow	Video recording	[182]
Neck stretching	Dairy cow	Video recording	[167]
	Dairy cow	Video recording	[179]
Positive bias	Sheep	Direct observations	[152]
Proportion of right-lateralised ears	Sheep	Direct observations	[158]
Qualitative Behaviour Assessment	Beef cattle	Direct observations	[162]
	Beef cattle	Direct observations	[165]
	Dairy cow	Direct observations	[162]
	Goat	Direct observations	[166]
	Sheep	Video recording	[163]
	Sheep	Direct observations	[164]
	Veal calf	Direct observations	[162]
Rubbing stroker	Dairy cow	Video recording	[167]
Ruminating	Sheep	Video recording	[173]
Sniffing stroker	Dairy cow	Video recording	[167]
Tail up	Goat	Video recording	[181]
Tail wagging	Sheep	Video recording	[173]
Total duration of tail wagging	Sheep	Video recording	[172]
Vigorous tail wagging	Dairy cow	Video recording	[182]

4. Conclusions

The present review allowed the identification of a list of promising indicators that might be included in welfare assessment protocols for ruminants. Most of them cover several aspects of positive welfare related to three domains: Environment, Behaviour, and Mental State. Few positive welfare indicators are available for the evaluation of Nutrition and Health. Thus, further research is needed for these last domains.

Many indicators can be considered valid to highlight positive welfare conditions. However, much of the validity evidence is based on their absence when a negative situation is present; in order to affirm their validity as indicators of positive welfare, it would, therefore, be important to validate them also in the opposite direction, demonstrating their increase in pleasurable situations. This should be a relevant topic for future research. Reliability also needs to be further investigated, as very few studies focus on this aspect; however, in the very few cases where it has been tested, the results seem to be good.

Some indicators are apparently feasible on-farm, but most of them require the development of specific sampling strategies and/or rely on the use of video- or automatic-recording devices. Automation and the use of machine learning systems would help to increase feasibility.

In conclusion, several indicators are potentially already available (e.g., synchronisation of lying and feeding, coat or fleece condition, QBA), but further indicators are required for some domains, and some further testing and refinement is needed for those that are already available. Filling these gaps would be extremely useful in order to set up new welfare assessment protocols able to focus on indicators of positive welfare, in order to assure a higher quality of animal products to consumers, with the guarantee that animals are really experiencing a life that is worth living.

Author Contributions: Conceptualisation, S.M. and M.B.; writing—original draft preparation, S.M., M.B., G.D.R., F.N.; writing—review and editing, C.D.

Funding: This review received no external funding.

Acknowledgments: No financial support nor external contribution was received for writing the present review.

Conflicts of Interest: The authors declare no conflict of interest.

References

1. Welfare Quality Consortium. *Welfare Quality® Assessment Protocol for Cattle*; Welfare Quality Consortium: Lelystad, The Netherlands, 2009.
2. Battini, M.; Stilwell, G.; Vieira, A.; Barbieri, S.; Canali, E.; Mattiello, S. On-farm welfare assessment protocol for adult dairy goats in intensive production systems. *Animals* **2015**, *5*, 934–950. [CrossRef] [PubMed]
3. AWIN. *AWIN Welfare Assessment Protocol for Goats*; AWIN: Berlin, Germany, 2015; p. 70. [CrossRef]
4. Caroprese, M.; Napolitano, F.; Mattiello, S.; Fthenakis, G.C.; Ribó, O.; Sevi, A. On-farm welfare monitoring of small ruminants. *Small Rumin. Res.* **2016**, *135*, 20–25. [CrossRef]
5. AWIN. *AWIN Welfare Assessment Protocol for Sheep*; AWIN: Berlin, Germany, 2015; p. 69. [CrossRef]
6. EFSA. Statement on the use of animal-based measures to assess the welfare of animals. *EFSA J.* **2012**, *10*, 1–29.
7. Brambell Report. *Report of the Technical Committee to Enquire into the Welfare of Animal Kept under Intensive Livestock Husbandry Systems*; Brambell Report: London, UK, 1965.
8. Yeates, J.W.; Main, D.C.J. Assessment of positive welfare: A review. *Vet. J.* **2008**, *175*, 293–300. [CrossRef] [PubMed]
9. Farm Animal Welfare Council. *Farm Animal Welfare in Great Britain: Past, Present and Future*; Farm Animal Welfare Council: London, UK, 2009.
10. Fraser, D. Understanding animal welfare. *Acta Vet. Scand.* **2008**, *50*. [CrossRef]
11. Burow, E.; Rousing, T.; Thomsen, P.T.; Otten, N.D.; Sorensen, J.T. Effect of grazing on the cow welfare of dairy herds evaluated by a multidimensional welfare index. *Animal* **2013**, *7*, 834–842. [CrossRef]
12. Sachser, N. What is important to achieve good welfare in animals? In *Dahlem Workshop Report 87—Coping with Challenge—Welfare in Animals Including Humans*; Broom, D.M., Ed.; Dahlem University Press: Berlin, Germany, 2001; pp. 31–48.
13. Mellor, D.J. Enhancing animal welfare by creating opportunities for positive affective engagement. *N. Z. Vet. J.* **2015**, *63*, 3–8. [CrossRef]
14. Green, T.C.; Mellor, D.J. Extending ideas about animal welfare assessment to include "quality of life" and related concepts. *N. Z. Vet. J.* **2011**, *59*, 263–271. [CrossRef]
15. Mellor, D.J. Updating animal welfare thinking: Moving beyond the "five freedoms" towards "A life worth living". *Animals* **2016**, *6*, 21. [CrossRef]
16. OIE. Introduction to the recommendations for animal welfare. *Terr. Anim. Heal. Code* **2019**, *1*, 1–4.
17. Vigors, B. Citizens' and Farmers' Framing of 'Positive Animal Welfare' and the Implications for Framing Positive Welfare in Communication. *Animals* **2019**, *9*, 147. [CrossRef] [PubMed]
18. Mellor, D.J. Operational details of the five domains model and its key applications to the assessment and management of animal welfare. *Animals* **2017**, *7*, 60. [CrossRef] [PubMed]
19. Mellor, D.J.; Beausoleil, N.J. Extending the "Five Domains" model for animal welfare assessment to incorporate positive welfare states. *Anim. Welf.* **2015**, *24*, 241–253. [CrossRef]
20. Battini, M.; Vieira, A.; Barbieri, S.; Ajuda, I.; Stilwell, G.; Mattiello, S. Invited review: Animal-based indicators for on-farm welfare assessment for dairy goats. *J. Dairy Sci.* **2014**, *97*, 6625–6648. [CrossRef]
21. Verbeek, E.; Ferguson, D.; Quinquet de Monjour, P.; Lee, C. Generating positive affective states in sheep: The influence of food rewards and opioid administration. *Appl. Anim. Behav. Sci.* **2014**, *154*, 39–47. [CrossRef]
22. Kilgour, R.J.; Uetake, K.; Ishiwata, T.; Melville, G.J. The behaviour of beef cattle at pasture. *Appl. Anim. Behav. Sci.* **2012**, *138*, 12–17. [CrossRef]

23. Tuomisto, L.; Huuskonen, A.; Jauhiainen, L.; Mononen, J. Finishing bulls have more synchronised behaviour in pastures than in pens. *Appl. Anim. Behav. Sci.* **2019**, *213*, 26–32. [CrossRef]
24. Tölü, C.; Göktürk, S.; Savaş, T. Effects of weaning and spatial enrichment on behavior of Turkish saanen goat kids. *Asian Australas. J. Anim. Sci.* **2016**, *29*, 879–886. [CrossRef]
25. Stachowicz, J.; Gygax, L.; Hillmann, E.; Wechsler, B.; Keil, N.M. Dairy goats use outdoor runs of high quality more regardless of the quality of indoor housing. *Appl. Anim. Behav. Sci.* **2018**, *208*, 22–30. [CrossRef]
26. Overton, M.W.; Moore, D.A.; Sischo, W.M. Comparison of commonly used ndices to evaluate dairy cattle lying behavior. In *Proceedings of the Fifth International Dairy Housing Proceedings*; Janni, K., Ed.; ASAE Publication Number 701P0203; ASAE: Fort Worth, TX, USA, 2003.
27. Haley, D.B.; Rushen, J.; de Passillé, A.M. Behavioural indicators of cow comfort: Activity and resting behaviour of dairy cows in two types of housing. *Can. J. Anim. Sci.* **2010**, *80*, 257–263. [CrossRef]
28. Drissler, M.; Gaworski, M.; Tucker, C.B.; Weary, D.M. Freestall Maintenance: Effects on Lying Behavior of Dairy Cattle. *J. Dairy Sci.* **2010**, *88*, 2381–2387. [CrossRef]
29. Sutherland, M.A.; Stewart, M.; Schütz, K.E. Effects of two substrate types on the behaviour, cleanliness and thermoregulation of dairy calves. *Appl. Anim. Behav. Sci.* **2013**, *147*, 19–27. [CrossRef]
30. Sahu, D.; Mandal, D.K.; Hussain Dar, A.; Podder, M.; Gupta, A. Modification in housing system affects the behavior and welfare of dairy Jersey crossbred cows in different seasons. *Biol. Rhythm Res.* **2019**. [CrossRef]
31. Norring, M.; Manninen, E.; de Passillé, A.M.; Rushen, J.; Saloniemi, H. Preferences of dairy cows for three stall surface materials with small amounts of bedding. *J. Dairy Sci.* **2010**, *93*, 70–74. [CrossRef] [PubMed]
32. Ehrlenbruch, R.; Jørgensen, G.H.M.; Andersen, I.L.; Bøe, K.E. Provision of additional walls in the resting area—The effects on resting behaviour and social interactions in goats. *Appl. Anim. Behav. Sci.* **2010**, *122*, 35–40. [CrossRef]
33. Nielsen, L.H.; Mogensen, L.; Krohn, C.; Hindhede, J.; Sørensen, J.T. Resting and social behaviour of dairy heifers housed in slatted floor pens with different sized bedded lying areas. *Appl. Anim. Behav. Sci.* **1997**, *54*, 307–316. [CrossRef]
34. Jørgensen, G.H.M.; Andersen, I.L.; Bøe, K.E. The effect of different pen partition configurations on the behaviour of sheep. *Appl. Anim. Behav. Sci.* **2009**, *119*, 66–70. [CrossRef]
35. Færevik, G.; Andersen, I.L.; Bøe, K.E. Preferences of sheep for different types of pen flooring. *Appl. Anim. Behav. Sci.* **2005**, *90*, 265–276. [CrossRef]
36. Bøe, K.E.; Ehrlenbruch, R.; Andersen, I.L. Outside enclosure and additional enrichment for dairy goats—A preliminary study. *Acta Vet. Scand.* **2012**, *54*, 68. [CrossRef]
37. Platz, S.; Ahrens, F.; Bendel, J.; Meyer, H.H.D.; Erhard, M.H. What Happens with Cow Behavior When Replacing Concrete Slatted Floor by Rubber Coating: A Case Study. *J. Dairy Sci.* **2008**, *91*, 999–1004. [CrossRef]
38. Hörning, B. Attempts to integrate different parameters into an overall picture of animal welfare using investigations in dairy loose houses as an example. *Anim. Welf.* **2003**, *12*, 557–563.
39. Plesch, G.; Broerkens, N.; Laister, S.; Winckler, C.; Knierim, U. Reliability and feasibility of selected measures concerning resting behaviour for the on-farm welfare assessment in dairy cows. *Appl. Anim. Behav. Sci.* **2010**, *126*, 19–26. [CrossRef]
40. Van Erp-van der Kooij, E.; Almalik, O.; Cavestany, D.; Roelofs, J.; van Eerdenburg, F. Lying Postures of Dairy Cows in Cubicles and on Pasture. *Animals* **2019**, *9*, 183. [CrossRef] [PubMed]
41. Krohn, C.C.; Munksgaard, L.; Jonasen, B. Behaviour of dairy cows kept in extensive (loose housing/pasture) or intensive (tie stall) environments. I. Experimental procedures, facilities, time budgets-diurnal and seasonal conditions. *Appl. Anim. Behav. Sci.* **1992**, *34*, 37–47. [CrossRef]
42. Battini, M.; Peric, T.; Ajuda, I.; Vieira, A.; Grosso, L.; Barbieri, S.; Stilwell, G.; Prandi, A.; Comin, A.; Tubaro, F.; et al. Hair coat condition: A valid and reliable indicator for on-farm welfare assessment in adult dairy goats. *Small Rumin. Res.* **2015**, *123*, 197–203. [CrossRef]
43. De Vries, A. Economic trade-offs between genetic improvement and longevity in dairy cattle. *J. Dairy Sci.* **2017**, *100*, 4184–4192. [CrossRef]
44. Matheson, S.M.; Rooke, J.A.; McIlvaney, K.; Jack, M.; Ison, S.; Bnger, L.; Dwyer, C.M. Development and validation of on-farm behavioural scoring systems to assess birth assistance and lamb vigour. *Animal* **2011**, *5*, 776–783. [CrossRef]

45. Matheson, S.M.; Bünger, L.; Dwyer, C.M. Genetic parameters for fitness and neonatal behavior traits in sheep. *Behav. Genet.* **2012**, *42*, 899–911. [CrossRef]
46. Val-Laillet, D.; Guesdon, V.; von Keyserlingk, M.A.G.; de Passillé, A.M.; Rushen, J. Allogrooming in cattle: Relationships between social preferences, feeding displacements and social dominance. *Appl. Anim. Behav. Sci.* **2009**, *116*, 141–149. [CrossRef]
47. Gutmann, A.K.; Špinka, M.; Winckler, C. Long-term familiarity creates preferred social partners in dairy cows. *Appl. Anim. Behav. Sci.* **2015**, *169*, 1–8. [CrossRef]
48. Windschnurer, I.; Schmied, C.; Boivin, X.; Waiblinger, S. Assessment of Human-Animal Relationships in Dairy Cows. In *Welfare Quality® Reports*; Forkman, B., Keeling, L., Eds.; Cardiff University: Cardiff, UK, 2009; Volume 11, pp. 137–152.
49. Napolitano, F.; Serrapica, F.; Braghieri, A.; Masucci, F.; Sabia, E.; De Rosa, G. Human-Animal Interactions in Dairy Buffalo Farms. *Animals* **2019**, *9*, 246. [CrossRef] [PubMed]
50. Windschnurer, I.; Schmied, C.; Boivin, X.; Waiblinger, S. Reliability and inter-test relationship of tests for on-farm assessment of dairy cows' relationship to humans. *Appl. Anim. Behav. Sci.* **2008**, *114*, 37–53. [CrossRef]
51. Waiblinger, S.; Menke, C.; Coleman, G. The relationship between attitudes, personal characteristics and behaviour of stockpeople and subsequent behaviour and production of dairy cows. *Appl. Anim. Behav. Sci.* **2002**, *79*, 195–219. [CrossRef]
52. Westerath, H.S.; Laister, S.; Winckler, C.; Knierim, U. Exploration as an indicator of good welfare in beef bulls: An attempt to develop a test for on-farm assessment. *Appl. Anim. Behav. Sci.* **2009**, *116*, 126–133. [CrossRef]
53. Mintline, E.M.; Wood, S.L.; de Passillé, A.M.; Rushen, J.; Tucker, C.B. Assessing calf play behavior in an arena test. *Appl. Anim. Behav. Sci.* **2012**, *141*, 101–107. [CrossRef]
54. De Rosa, G.; Grasso, F.; Braghieri, A.; Bilancione, A.; Di Francia, A.; Napolitano, F. Behavior and milk production of buffalo cows as affected by housing system. *J. Dairy Sci.* **2009**. [CrossRef]
55. Ninomiya, S. Grooming Device Effects on Behaviour and Welfare of Japanese Black Fattening Cattle. *Animals* **2019**, *9*, 186. [CrossRef]
56. Westerath, H.S.; Gygax, L.; Hillmann, E. Are special feed and being brushed judged as positive by calves? *Appl. Anim. Behav. Sci.* **2014**. [CrossRef]
57. Favreau-Peigné, A.; Baumont, R.; Ginane, C. Food sensory characteristics: Their unconsidered roles in the feeding behaviour of domestic ruminants. *Animal* **2013**, *7*, 806–813. [CrossRef]
58. Manteca, X.; Villalba, J.J.; Atwood, S.B.; Dziba, L.; Provenza, F.D. Is dietary choice important to animal welfare? *J. Vet. Behav. Clin. Appl. Res.* **2008**, *3*, 229–239. [CrossRef]
59. Rutter, S.M. Review: Grazing preferences in sheep and cattle: Implications for production, the environment and animal welfare. *Can. J. Anim. Sci.* **2010**, *90*, 285–293. [CrossRef]
60. De Rosa, G.; Napolitano, F.; Marino, V.; Bordi, A. Induction of conditioned taste aversion in goats. *Small Rumin. Res.* **1995**, *16*, 7–11. [CrossRef]
61. Provenza, F. Postingestive Feedback as an Elementary Determinant of Food Preference and Intake in Ruminants. *J. Range Manag.* **1995**, *48*, 2–17. [CrossRef]
62. Catanese, F.; Obelar, M.; Villalba, J.J.; Distel, R.A. The importance of diet choice on stress-related responses by lambs. *Appl. Anim. Behav. Sci.* **2013**, *148*, 37–45. [CrossRef]
63. Dubé, L.; LeBel, J.L.; Lu, J. Affect asymmetry and comfort food consumption. *Physiol. Behav.* **2005**, *86*, 559–567. [CrossRef]
64. Lin, J.Y.; Amodeo, L.R.; Arthurs, J.; Reilly, S. Taste neophobia and palatability: The pleasure of drinking. *Physiol. Behav.* **2012**, *106*, 515–519. [CrossRef]
65. Webb, L.E.; Engel, B.; Berends, H.; van Reenen, C.G.; Gerrits, W.J.J.; de Boer, I.J.M.; Bokkers, E.A.M. What do calves choose to eat and how do preferences affect behaviour? *Appl. Anim. Behav. Sci.* **2014**, *161*, 7–19. [CrossRef]
66. Meagher, R.K.; Weary, D.M.; von Keyserlingk, M.A.G. Some like it varied: Individual differences in preference for feed variety in dairy heifers. *Appl. Anim. Behav. Sci.* **2017**, *195*, 8–14. [CrossRef]
67. Atwood, S.B.; Provenza, F.D.; Wiedmeier, R.D.; Banner, R.E. Influence of free-choice vs mixed-ration diets on food intake and performance of fattening calves. *J. Anim. Sci.* **2001**, *79*, 3034–3040. [CrossRef]
68. Napolitano, F.; Knierim, U.; Grass, F.; De Rosa, G. Positive indicators of cattle welfare and their applicability to on-farm protocols. *Ital. J. Anim. Sci.* **2010**, *8*, 355–365. [CrossRef]

69. Miranda-de la Lama, G.C.; Mattiello, S. The importance of social behaviour for goat welfare in livestock farming. *Small Rumin. Res.* **2010**, *90*, 1–10. [CrossRef]
70. Gautrais, J.; Michelena, P.; Sibbald, A.; Bon, R.; Deneubourg, J.L. Allelomimetic synchronization in Merino sheep. *Anim. Behav.* **2007**, *74*, 1443–1454. [CrossRef]
71. Bouissou, M.F.; Boissy, A.; Le Neindre, P.; Veissier, I. The social behaviour of cattle. In *Social Behaviour in Farm Animals*; Keeling, L., Gonyou, H., Eds.; CAB International: Wallingford, UK, 2001; pp. 113–145. ISBN 0-85199-397-4.
72. Dávid-Barrett, T.; Dunbar, R.I.M. Cooperation, behavioural synchrony and status in social networks. *J. Theor. Biol.* **2012**, *308*, 88–95. [CrossRef] [PubMed]
73. Stoye, S.; Porter, M.A.; Stamp Dawkins, M. Synchronized lying in cattle in relation to time of day. *Livest. Sci.* **2012**, *149*, 70–73. [CrossRef]
74. Muñoz-Osorio, G.A.; Aguilar-Caballero, A.J.; Cámara-Sarmiento, R. Influencia del tipo de alojamiento sobre el comportamiento productivo y bienestar de corderos en sistemas de engorda intensivos. *Trop. Subtrop. Agroecosyst.* **2019**, *22*, 1–11.
75. Petherick, J.C.; Phillips, C.J.C. Space allowances for confined livestock and their determination from allometric principles. *Appl. Anim. Behav. Sci.* **2009**, *117*, 1–12. [CrossRef]
76. Mandel, R.; Whay, H.R.; Klement, E.; Nicol, C.J. Invited review: Environmental enrichment of dairy cows and calves in indoor housing. *J. Dairy Sci.* **2016**, *99*, 1695–1715. [CrossRef]
77. Krohn, C.C.; Munksgaard, L. Krohn & Munksgaard, 1993_lying in cattle.pdf. *Appl. Anim. Behav. Sci.* **1993**, *37*, 1–16.
78. Lidfors, L. The use of getting up and lying down movements in the evaluation of cattle environments. *Vet. Res. Commun.* **1989**, *13*, 307–324. [CrossRef]
79. Jensen, M.B.; Pedersen, L.J.; Munksgaard, L. The effect of reward duration on demand functions for rest in dairy heifers and lying requirements as measured by demand functions. *Appl. Anim. Behav. Sci.* **2005**, *90*, 207–217. [CrossRef]
80. Hansen, I. Behavioural indicators of sheep and goat welfare in organic and conventional Norwegian farms. *Acta Agric. Scand. A Anim. Sci.* **2015**, *65*, 55–61. [CrossRef]
81. Richmond, S.E.; Wemelsfelder, F.; de Heredia, I.B.; Ruiz, R.; Canali, E.; Dwyer, C.M. Evaluation of Animal-Based Indicators to Be Used in a Welfare Assessment Protocol for Sheep. *Front. Vet. Sci.* **2017**, *4*, 1–13. [CrossRef] [PubMed]
82. Phillips, C.J.C. *Cattle Behaviour and Welfare*, 2nd ed.; Wiley-Blackwell Science Ltd.: Oxford, UK, 2002.
83. Napolitano, F.; Pacelli, C.; De Rosa, G.; Braghieri, A.; Girolami, A. Sustainability and welfare of Podolian cattle. *Livest. Prod. Sci.* **2005**, *92*, 323–331. [CrossRef]
84. Mattiello, S.; Battini, M.; Andreoli, E.; Barbieri, S. Short communication: Breed differences affecting dairy cattle welfare in traditional alpine tie-stall husbandry systems. *J. Dairy Sci.* **2011**, *94*. [CrossRef] [PubMed]
85. Franco, N.H.; Magalhães-Sant'Ana, M.; Olsson, I.A.S. Welfare and quantity of life. In *Dilemmas in Animal Welfare*; Appleby, M., Sandøe, P., Weary, D., Eds.; CABI: Wallingford, UK, 2014; pp. 46–66.
86. Can, E.; Vieira, A.; Battini, M.; Mattiello, S.; Stilwell, G. Consistency over time of animal-based welfare indicators as a further step for developing a welfare assessment monitoring scheme: The case of the Animal Welfare Indicators protocol for dairy goats. *J. Dairy Sci.* **2017**, *100*, 9194–9204. [CrossRef] [PubMed]
87. Phythian, C.J.; Cripps, P.J.; Michalopoulou, E.; Jones, P.H.; Grove-White, D.; Clarkson, M.J.; Winter, A.C.; Stubbings, L.A.; Duncan, J.S. Reliability of indicators of sheep welfare assessed by a group observation method. *Vet. J.* **2012**, *193*, 257–263. [CrossRef]
88. Mellor, D.J. Positive animal welfare states and encouraging environment-focused and animal-to-animal interactive behaviours. *N. Z. Vet. J.* **2015**, *63*, 9–16. [CrossRef]
89. Krachun, C.; Rushen, J.; de Passillé, A.M. Play behaviour in dairy calves is reduced by weaning and by a low energy intake. *Appl. Anim. Behav. Sci.* **2010**, *122*, 71–76. [CrossRef]
90. Thornton, P.D.; Waterman-Pearson, A.E. Behavioural responses to castration in lambs. *Anim. Welf.* **2002**, *11*, 203–212.
91. Burghardt, G. The genesis of animal play. *Nature* **2005**, *434*, 273.
92. Boissy, A.; Manteuffel, G.; Jensen, M.B.; Moe, R.O.; Spruijt, B.; Keeling, L.J.; Winckler, C.; Forkman, B.; Dimitrov, I.; Langbein, J.; et al. Assessment of positive emotions in animals to improve their welfare. *Physiol. Behav.* **2007**, *92*, 375–397. [CrossRef] [PubMed]

93. Valníčková, B.; Stěhulová, I.; Šárová, R.; Špinka, M. The effect of age at separation from the dam and presence of social companions on play behavior and weight gain in dairy calves. *J. Dairy Sci.* **2015**, *98*, 5545–5556. [CrossRef] [PubMed]
94. Loberg, J.; Telezhenko, E.; Bergsten, C.; Lidfors, L. Behaviour and claw health in tied dairy cows with varying access to exercise in an outdoor paddock. *Appl. Anim. Behav. Sci.* **2004**, *89*, 1–16. [CrossRef]
95. Anderson, C.; Yngvesson, J.; Boissy, A.; Uvnäs-Moberg, K.; Lidfors, L. Behavioural expression of positive anticipation for food or opportunity to play in lambs. *Behav. Process.* **2015**, *113*, 152–158. [CrossRef]
96. Moe, R.O.; Nordgreen, J.; Janczak, A.M.; Spruijt, B.M.; Zanella, A.J.; Bakken, M. Trace classical conditioning as an approach to the study of reward-related behaviour in laying hens: A methodological study. *Appl. Anim. Behav. Sci.* **2009**, *121*, 171–178. [CrossRef]
97. Held, S.D.E.; Špinka, M. Animal play and animal welfare. *Anim. Behav.* **2011**, *81*, 891–899. [CrossRef]
98. Gygax, L.; Hillmann, E. "Naturalness" and Its Relation to Animal Welfare from an Ethological Perspective. *Agriculture* **2018**, *8*, 136. [CrossRef]
99. Mattiello, S.; Ferrante, V.; Verga, M.; Gottardo, F.; Andrighetto, I.; Canali, E.; Caniatti, M.; Cozzi, G. The provision of solid feeds to veal calves: II. Behavior, physiology, and abomasal damage1. *J. Anim. Sci.* **2016**, *80*, 367–375. [CrossRef]
100. Napolitano, F.; Annicchiarico, G.; Caroprese, M.; De Rosa, G.; Taibi, L.; Sevi, A. Lambs prevented from suckling their mothers display behavioral, immune and endocrine disturbances. *Physiol. Behav.* **2003**, *78*, 81–89. [CrossRef]
101. Arnold, G.W.; Dudzinski, M.L. Social organization and animal dispersion. In *Ethology of Free-Ranging Domestic Animals*; Arnold, G.W., Dudzinski, M.L., Eds.; Elsevier Scientific Publishing Company: Amsterdam, Switzerland, 1978; pp. 51–96.
102. Arnott, G.; Ferris, C.P.; O'connell, N.E. Review: Welfare of dairy cows in continuously housed and pasture-based production systems. *Animal* **2017**, *11*, 261–273. [CrossRef]
103. Sato, S.; Tarumizu, K.; Hatae, K. The influence of social factors on allogrooming in cows. *Appl. Anim. Behav. Sci.* **1993**, *38*, 235–244. [CrossRef]
104. Sato, S.; Sako, S.; Maeda, A. Social licking patterns in cattle (Bos taurus): Influence of environmental and social factors. *Appl. Anim. Behav. Sci.* **1991**, *32*, 3–12. [CrossRef]
105. Baxter, E.M.; Mulligan, J.; Hall, S.A.; Donbavand, J.E.; Palme, R.; Aldujaili, E.; Zanella, A.J.; Dwyer, C.M. Positive and negative gestational handling influences placental traits and mother-offspring behavior in dairy goats. *Physiol. Behav.* **2016**, *157*, 129–138. [CrossRef] [PubMed]
106. Laister, S.; Stockinger, B.; Regner, A.M.; Zenger, K.; Knierim, U.; Winckler, C. Social licking in dairy cattle-Effects on heart rate in performers and receivers. *Appl. Anim. Behav. Sci.* **2011**, *130*, 81–90. [CrossRef]
107. Wood, M.T. Social grooming patterns in two herds of monozygotic twin dairy cows. *Anim. Behav.* **1977**, *25*, 635–642. [CrossRef]
108. Sato, S. Social licking pattern and its relationships to social dominance and live weight gain in weaned calves. *Appl. Anim. Behav. Sci.* **1984**, *12*, 25–32. [CrossRef]
109. Tresoldi, G.; Weary, D.M.; Filho, L.C.P.M.; von Keyserlingk, M.A.G. Social licking in pregnant dairy heifers. *Animals* **2015**, *5*, 1169–1179. [CrossRef]
110. Krohn, C.C. Behaviour of dairy cows kept in extensive(loose housing/pasture) or intensive (tie stall) environments. III. Grooming, exploration and abnormal behaviour. *Appl. Anim. Behav. Sci.* **1994**, *42*, 73–86. [CrossRef]
111. Jensen, M.B.; Herskin, M.S.; Thomsen, P.T.; Forkman, B.; Houe, H. Preferences of lame cows for type of surface and level of social contact in hospital pens. *J. Dairy Sci.* **2015**, *98*, 4552–4559. [CrossRef]
112. Mooring, M.S.; Gavazzi, A.J.; Hart, B.L. Effects of castration on grooming in goats. *Physiol. Behav.* **1998**, *64*, 707–713. [CrossRef]
113. Kakuma, Y.; Takeuchi, Y.; Mori, Y.; Hart, B.L. Hormonal control of grooming behavior in domestic goats. *Physiol. Behav.* **2003**, *78*, 61–66. [CrossRef]
114. Hart, B.L.; Pryor, P.A. Developmental and hair-coat determinants of grooming behaviour in goats and sheep. *Anim. Behav.* **2004**, *67*, 11–19. [CrossRef]
115. Mooring, M.S.; Hart, B.L.; Fitzpatrick, T.A.; Reisig, D.D.; Nishihira, T.T.; Fraser, I.C.; Benjamin, J.E. Grooming in desert bighorn sheep (*Ovis canadensis mexicana*) and the ghost of parasites past. *Behav. Ecol.* **2006**, *17*, 364–371. [CrossRef]

116. Wilson, L.L.; Terosky, T.L.; Stull, C.L.; Stricklin, W.R. Effects of Individual Housing Design and Size on Behavior and Stress Indicators of Special-Fed Holstein Veal Calves. *J. Anim. Sci.* **1999**, *77*, 1341–1347. [CrossRef]
117. Rushen, J.; de Passillé, A.M.B. The scientific assessment of the impact of housing on animal welfare: A critical review. *Can. J. Anim. Sci.* **1992**, *72*, 721–743. [CrossRef]
118. Broom, D.M. Needs and welfare of housed calves. In *New Trends in Veal Calf Production*; Metz, J.M., Groenestein, C.M., Eds.; EAAP Publication n. 52: Pudoc; EAAP: Wageningen, The Netherlands, 1991; pp. 23–31.
119. McConnachie, E.; Smid, A.M.C.; Thompson, A.J.; Weary, D.M.; Gaworski, M.A.; Von Keyserlingk, M.A.G. Cows are highly motivated to access a grooming substrate. *Biol. Lett.* **2018**, *14*, 1–4. [CrossRef]
120. Mandel, R.; Whay, H.R.; Nicol, C.J.; Klement, E. The effect of food location, heat load, and intrusive medical procedures on brushing activity in dairy cows. *J. Dairy Sci.* **2013**, *96*, 6506–6513. [CrossRef]
121. De Vries, M.; Bokkers, E.A.M.; van Reenen, C.G.; Engel, B.; van Schaik, G.; Dijkstra, T.; de Boer, I.J.M. Housing and management factors associated with indicators of dairy cattle welfare. *Prev. Vet. Med.* **2015**, *118*, 80–92. [CrossRef]
122. Bertenshaw, C.; Rowlinson, P.; Edge, H.; Douglas, S.; Shiel, R. The effect of different degrees of "positive" human-animal interaction during rearing on the welfare and subsequent production of commercial dairy heifers. *Appl. Anim. Behav. Sci.* **2008**, *114*, 65–75. [CrossRef]
123. Sabia, E.; Napolitano, F.; De Rosa, G.; Terzano, G.M.; Barile, V.L.; Braghieri, A.; Pacelli, C. Efficiency to reach age of puberty and behaviour of buffalo heifers (*Bubalus bubalis*) kept on pasture or in confinement. *Animal* **2014**, *8*, 1907–1916. [CrossRef]
124. Hemsworth, P.H. Human-animal interactions in livestock production. *Appl. Anim. Behav. Sci.* **2003**, *81*, 185–198. [CrossRef]
125. Breuer, K.; Hemsworth, P.; Barnett, J.; Matthews, L.; Coleman, G. Behavioural response to humans and the productivity of commercial dairy cows. *Appl. Anim. Behav. Sci.* **2000**, *66*, 273–288. [CrossRef]
126. Lensink, B.J.; Fernandez, X.; Boivin, X.; Pradel, P. The impact of gentle contacts on ease of handling, welfare, and. *J. Anim. Sci.* **2000**, *78*, 1219–1226. [CrossRef] [PubMed]
127. Rushen, J.; de Passillé, A.M.B.; Munksgaard, L. Fear of People by Cows and Effects on Milk Yield, Behavior, and Heart Rate at Milking. *J. Dairy Sci.* **2010**, *82*, 720–727. [CrossRef]
128. Hemsworth, P.H.; Coleman, G.J.; Barnett, J.L.; Borg, S. Relationships between human-animal interactions and productivity of commercial dairy cows. *J. Anim. Sci.* **2000**, *78*, 2821–2831. [CrossRef] [PubMed]
129. Lürzel, S.; Münsch, C.; Windschnurer, I.; Futschik, A.; Palme, R.; Waiblinger, S. The influence of gentle interactions on avoidance distance towards humans, weight gain and physiological parameters in group-housed dairy calves. *Appl. Anim. Behav. Sci.* **2015**, *172*, 9–16. [CrossRef]
130. Serrapica, M.; Boivin, X.; Coulon, M.; Braghieri, A.; Napolitano, F. Positive perception of human stroking by lambs: Qualitative behaviour assessment confirms previous interpretation of quantitative data. *Appl. Anim. Behav. Sci.* **2017**, *187*, 31–37. [CrossRef]
131. Coulon, M.; Nowak, R.; Peyrat, J.; Chandèze, H.; Boissy, A.; Boivin, X. Do Lambs Perceive Regular Human Stroking as Pleasant? Behavior and Heart Rate Variability Analyses. *PLoS ONE* **2015**, *10*, e0118617. [CrossRef]
132. Guesdon, V.; Nowak, R.; Meurisse, M.; Boivin, X.; Cornilleau, F.; Chaillou, E.; Lévy, F. Behavioral evidence of heterospecific bonding between the lamb and the human caregiver and mapping of associated brain network. *Psychoneuroendocrinology* **2016**, *71*, 159–169. [CrossRef]
133. Ellingsen, K.; Coleman, G.J.; Lund, V.; Mejdell, C.M. Using qualitative behaviour assessment to explore the link between stockperson behaviour and dairy calf behaviour. *Appl. Anim. Behav. Sci.* **2014**, *153*, 10–17. [CrossRef]
134. Winckler, C.; Brinkmann, J.; Glatz, J. Long-term consistency of selected animal-related welfare parameters in dairy farms. *Anim. Welf.* **2007**, *16*, 197–199.
135. Waiblinger, S.; Menke, C.; Fölsch, D.W. Influences on the avoidance and approach behaviour of dairy cows towards humans on 35 farms. *Appl. Anim. Behav. Sci.* **2003**, *84*, 23–39. [CrossRef]
136. Battini, M.; Barbieri, S.; Waiblinger, S.; Mattiello, S. Validity and feasibility of Human-Animal Relationship tests for on-farm welfare assessment in dairy goats. *Appl. Anim. Behav. Sci.* **2016**, *178*, 32–39. [CrossRef]
137. Tripaldi, C.; De Rosa, G.; Grasso, F.; Terzano, G.M.; Napolitano, F. Housing system and welfare of buffalo (*Bubalus bubalis*) cows. *Anim. Sci.* **2004**, *78*, 477–483. [CrossRef]

138. Napolitano, F.; Pacelli, C.; Grasso, F.; Braghieri, A.; De Rosa, G. The behaviour and welfare of buffaloes (*Bubalus bubalis*) in modern dairy enterprises. *Animal* **2013**, *7*, 1704–1713. [CrossRef] [PubMed]
139. Aggarwal, A.; Singh, M. Changes in skin and rectal temperature in lactating buffaloes provided with showers and wallowing during hot-dry season. *Trop. Anim. Health Prod.* **2008**, *40*, 223–228. [CrossRef] [PubMed]
140. Bracke, M.B.M. Review of wallowing in pigs: Description of the behaviour and its motivational basis. *Appl. Anim. Behav. Sci.* **2011**, *132*, 1–13. [CrossRef]
141. Hafez, E.S.E.; Cairns, R.B.; Hulet, C.V.; Scott, J.P. The behaviour of sheep and goats. In *The Behaviour of Domestic Animals*; Hafez, E.S.E., Ed.; Balliére Tindall: London, UK, 1969; pp. 296–348.
142. Aschwanden, J.; Gygax, L.; Wechsler, B.; Keil, N.M. Loose housing of small goat groups: Influence of visual cover and elevated levels on feeding, resting and agonistic behaviour. *Appl. Anim. Behav. Sci.* **2009**, *119*, 171–179. [CrossRef]
143. Pickup, H.E.; Dwyer, C.M. Breed differences in the expression of maternal care at parturition persist throughout the lactation period in sheep. *Appl. Anim. Behav. Sci.* **2011**, *132*, 33–41. [CrossRef]
144. Nowak, R.; Boivin, X. Filial attachment in sheep: Similarities and differences between ewe-lamb and human-lamb relationships. *Appl. Anim. Behav. Sci.* **2015**, *164*, 12–28. [CrossRef]
145. Muir, E.; Donbavand, J.; Dwyer, C.M. Salivary oxytocin is associated with ewe-lamb contact but not suckling in lactating ewes. In Proceedings of the 53rd Congress of the International Society of Applied Ethology, Bergen, Norway, 5–9 August 2019; p. 255.
146. Panksepp, J. The basic emotional circuits of mammalian brains: Do animals have affective lives? *Neurosci. Biobehav. Rev.* **2011**, *35*, 1791–1804. [CrossRef] [PubMed]
147. Mendl, M.; Burman, O.H.P.; Parker, R.M.A.; Paul, E.S. Cognitive bias as an indicator of animal emotion and welfare: Emerging evidence and underlying mechanisms. *Appl. Anim. Behav. Sci.* **2009**, *118*, 161–181. [CrossRef]
148. Baciadonna, L.; McElligott, A.G. The use of judgement bias to assess welfare in farm livestock. *Anim. Welf.* **2015**, *24*, 81–91. [CrossRef]
149. Roelofs, S.; Boleij, H.; Nordquist, R.E.; van der Staay, F.J. Making Decisions under Ambiguity: Judgment Bias Tasks for Assessing Emotional State in Animals. *Front. Behav. Neurosci.* **2016**, *10*, 1–16. [CrossRef] [PubMed]
150. Brajon, S.; Laforest, J.P.; Schmitt, O.; Devillers, N. The way humans behave modulates the emotional state of piglets. *PLoS ONE* **2015**, *10*, 1–17. [CrossRef] [PubMed]
151. Zidar, J.; Campderrich, I.; Jansson, E.; Wichman, A.; Winberg, S.; Keeling, L.; Løvlie, H. Environmental complexity buffers against stress-induced negative judgement bias in female chickens. *Sci. Rep.* **2018**, *8*, 1–14. [CrossRef]
152. Doyle, R.E.; Fisher, A.D.; Hinch, G.N.; Boissy, A.; Lee, C. Release from restraint generates a positive judgement bias in sheep. *Appl. Anim. Behav. Sci.* **2010**. [CrossRef]
153. Sanger, M.E.; Doyle, R.E.; Hinch, G.N.; Lee, C. Sheep exhibit a positive judgement bias and stress-induced hyperthermia following shearing. *Appl. Anim. Behav. Sci.* **2011**, *131*, 94–103. [CrossRef]
154. Crump, A.; Arnott, G.; Bethell, E.J. Affect-driven attention biases as animal welfare indicators: Review and methods. *Animals* **2018**, *8*, 136. [CrossRef]
155. Lee, C.; Verbeek, E.; Doyle, R.; Bateson, M. Attention bias to threat indicates anxiety differences in sheep. *Biol. Lett.* **2016**, *12*. [CrossRef]
156. Lee, C.; Cafe, L.M.; Robinson, S.L.; Doyle, R.E.; Lea, J.M.; Small, A.H.; Colditz, I.G. Anxiety influences attention bias but not flight speed and crush score in beef cattle. *Appl. Anim. Behav. Sci.* **2018**, *205*, 210–215. [CrossRef]
157. Whittaker, A.L.; Marsh, L.E. The role of behavioural assessment in determining 'positive' affective states in animals. *CAB Rev. Perspect. Agric. Vet. Sci. Nutr. Nat. Resour.* **2019**, *14*. [CrossRef]
158. Reefmann, N.; Bütikofer Kaszàs, F.; Wechsler, B.; Gygax, L. Ear and tail postures as indicators of emotional valence in sheep. *Appl. Anim. Behav. Sci.* **2009**, *118*, 199–207. [CrossRef]
159. Mendl, M.; Burman, O.H.P.; Paul, E.S. An integrative and functional framework for the study of animal emotion and mood. *Proc. R. Soc. B Biol. Sci.* **2010**, *277*, 2895–2904. [CrossRef]
160. Wemelsfelder, F.; Hunter, T.E.A.; Mendl, M.T.; Lawrence, A.B. Assessing the "whole animal": A free choice profiling approach. *Anim. Behav.* **2001**, *62*, 209–220. [CrossRef]

161. Napolitano, F.; De Rosa, G.; Braghieri, A.; Grasso, F.; Bordi, A.; Wemelsfelder, F. The qualitative assessment of responsiveness to environmental challenge in horses and ponies. *Appl. Anim. Behav. Sci.* **2008**, *109*, 342–354. [CrossRef]
162. Wemelsfelder, F.; Millard, F.; De Rosa, G.; Napolitano, F. Qualitative behaviour assessment. In *Welfare Quality® Report No. 11—Assessment of Animal Welfare Measures for Dairy Cattle, Beef Bulls and Veal Calves*; Forkman, B., Keeling, L., Eds.; Cardiff University: Cardiff, UK, 2009; pp. 215–224.
163. Phythian, C.; Michalopoulou, E.; Duncan, J.; Wemelsfelder, F. Inter-observer reliability of Qualitative Behavioural Assessments of sheep. *Appl. Anim. Behav. Sci.* **2013**, *144*, 73–79. [CrossRef]
164. Phythian, C.J.; Michalopoulou, E.; Cripps, P.J.; Duncan, J.S.; Wemelsfelder, F. On-farm qualitative behaviour assessment in sheep: Repeated measurements across time, and association with physical indicators of flock health and welfare. *Appl. Anim. Behav. Sci.* **2016**, *175*, 23–31. [CrossRef]
165. Sant'Anna, A.C.; Paranhos da Costa, M.J.R. Validity and feasibility of qualitative behavior assessment for the evaluation of Nellore cattle temperament. *Livest. Sci.* **2013**, *157*, 254–262. [CrossRef]
166. Grosso, L.; Battini, M.; Wemelsfelder, F.; Barbieri, S.; Minero, M.; Dalla Costa, E.; Mattiello, S. On-farm Qualitative Behaviour Assessment of dairy goats in different housing conditions. *Appl. Anim. Behav. Sci.* **2016**, *180*, 51–57. [CrossRef]
167. Proctor, H.S.; Carder, G. Measuring positive emotions in cows: Do visible eye whites tell us anything? *Physiol. Behav.* **2015**, *147*, 1–6. [CrossRef] [PubMed]
168. Sandem, A.I.; Janczak, A.M.; Salte, R.; Braastad, B.O. The use of diazepam as a pharmacological validation of eye white as an indicator of emotional state in dairy cows. *Appl. Anim. Behav. Sci.* **2006**, *96*, 177–183. [CrossRef]
169. Sandem, A.I.; Braastad, B.O.; Bakken, M. Behaviour and percentage eye-white in cows waiting to be fed concentrate—A brief report. *Appl. Anim. Behav. Sci.* **2006**, *97*, 145–151. [CrossRef]
170. Sandem, A.-I.; Braastad, B.O. Effects of cow-calf separation on visible eye white and behaviour in dairy cows—A brief report. *Appl. Anim. Behav. Sci.* **2005**, *95*, 233–239. [CrossRef]
171. Lambert (Proctor), H.S.; Carder, G. Looking into the eyes of a cow: Can eye whites be used as a measure of emotional state? *Appl. Anim. Behav. Sci.* **2017**, *186*, 1–6. [CrossRef]
172. Tamioso, P.R.; Rucinque, D.S.; Taconeli, C.A.; da Silva, G.P.; Molento, C.F.M. Behavior and body surface temperature as welfare indicators in selected sheep regularly brushed by a familiar observer. *J. Vet. Behav. Clin. Appl. Res.* **2017**, *19*, 27–34. [CrossRef]
173. Tamioso, P.R.; Maiolino Molento, C.F.; Boivin, X.; Chandèze, H.; Andanson, S.; Delval, É.; Hazard, D.; da Silva, G.P.; Taconeli, C.A.; Boissy, A. Inducing positive emotions: Behavioural and cardiac responses to human and brushing in ewes selected for high vs low social reactivity. *Appl. Anim. Behav. Sci.* **2018**, *208*, 56–65. [CrossRef]
174. Reefmann, N.; Bütikofer, F.; Wechsler, B.; Gygax, L. Physiological expression of emotional reactions in sheep. *Physiol. Behav.* **2009**, *98*, 235–241. [CrossRef]
175. Reefmann, N.; Wechsler, B.; Gygax, L. Behavioural and physiological assessment of positive and negative emotion in sheep. *Anim. Behav.* **2009**, *78*, 651–659. [CrossRef]
176. Battini, M.; Agostini, A.; Mattiello, S. Understanding cows' emotions on farm: Are eye white and ear posture reliable indicators? *Animals* **2019**, *9*, 477. [CrossRef]
177. Bellegarde, L.G.A.; Haskell, M.J.; Duvaux-ponter, C.; Weiss, A.; Boissy, A.; Erhard, H.W. Face-based perception of emotions in dairy goats. *Appl. Anim. Behav. Sci.* **2017**, *193*, 51–59. [CrossRef]
178. Proctor, H.S.; Carder, G. Can ear postures reliably measure the positive emotional state of cows? *Appl. Anim. Behav. Sci.* **2014**, *161*, 20–27. [CrossRef]
179. Schmied, C.; Waiblinger, S.; Scharl, T.; Leisch, F.; Boivin, X. Stroking of different body regions by a human: Effects on behaviour and heart rate of dairy cows. *Appl. Anim. Behav. Sci.* **2008**, *109*, 25–38. [CrossRef]
180. Boissy, A.; Aubert, A.; Greiveldinger, L.; Delval, E.; Veissier, I. Cognitive sciences to relate ear postures to emotions in sheep. *Anim. Welf.* **2011**, *20*, 47–56.
181. Briefer, E.F.; Tettamanti, F.; McElligott, A.G. Emotions in goats: Mapping physiological, behavioural and vocal profiles. *Anim. Behav.* **2015**, *99*, 131–143. [CrossRef]
182. De Oliveira, D.; Keeling, L.J. Routine activities and emotion in the life of dairy cows: Integrating body language into an affective state framework. *PLoS ONE* **2018**, *13*, 1–16. [CrossRef]

183. Padilla de la Torre, M.; Briefer, E.F.; Reader, T.; McElligott, A.G. Acoustic analysis of cattle (*Bos taurus*) mother-offspring contact calls from a source-filter theory perspective. *Appl. Anim. Behav. Sci.* **2015**, *163*, 58–68. [CrossRef]
184. Fisher, A.; Matthews, L. The social behavior of sheep. In *Social Behavior in Farm Animals*; Keeling, L., Gonyou, H., Eds.; CAB International: Wallingford, UK, 2001; pp. 211–245.

 © 2019 by the authors. Licensee MDPI, Basel, Switzerland. This article is an open access article distributed under the terms and conditions of the Creative Commons Attribution (CC BY) license (http://creativecommons.org/licenses/by/4.0/).

Review

Effects of Environmental Enrichment on Pig Welfare—A Review

Dorota Godyń [1,*], Jacek Nowicki [2] and Piotr Herbut [3]

[1] Department of Production Systems and Environment, National Research Institute of Animal Production, 32-083 Balice n. Kraków, Poland
[2] Department of Swine and Small Animal Breeding, Faculty of Animal Sciences, University of Agriculture in Krakow, 30-059 Kraków, Poland; j.nowicki@ur.krakow.pl
[3] Department of Rural Building, Faculty of Environmental Engineering and Land Surveying, University of Agriculture in Krakow, 30-059 Kraków, Poland; p.herbut@ur.krakow.pl
* Correspondence: dgodyn80@gmail.com

Received: 24 April 2019; Accepted: 19 June 2019; Published: 22 June 2019

Simple Summary: The legislation regarding pig housing systems states that environmental enrichment needs to be provided for group-housed pigs. Moreover, the materials used for improving animal housing are categorised as optimal, suboptimal, and of marginal interest. Straw has been considered as the optimal solution for pig housing, however there are some limitations in using it in a large amount. Therefore, other materials, objects, and toys have been used as enrichment in pig maintenance. Understanding how various enrichments influence animal welfare seems to be an important key to elaborating the best methods of improvement. This review presents new literature references regarding environmental enrichment for suckling piglets, weaning piglets, and fattening pigs.

Abstract: Good husbandry conditions on farms is of key importance for assuring animal welfare. One of the most important legal documents regulating the rules of maintaining pigs is the Directive 2008/120/EC, which states that group-housed pigs should have access to litter or other materials that provide exploration and occupation. Released in 2016, the Commission Recommendation (EU) 2016/336 on the application of the Council Directive 2008/120/EC characterizes the various categories of materials that may be used to improve animal welfare. According to the document, straw is considered as an optimal material for pig housing, however, materials categorized as suboptimal (e.g., wood bark) and materials of marginal interest (e.g., plastic toys) are often used in practice and scientific research. As such, the aim of this paper is to review and systematize the current state of knowledge on the topic of the impact of environmental enrichment on pig welfare. This article raises mainly issues, such as the effectiveness of the use of various enrichment on the reduction of undesirable behavior—tail biting; aggression; and stereotypies at the pre-weaning, post-weaning, and fattening stage of pig production.

Keywords: enrichment; pigs; welfare

1. Introduction

In recent years, there has been a growing interest from consumers into food production. Taking into consideration products of animal origin, it may be stated that the public debate is focused not only on the good quality of such products. Currently, a great deal of attention has also been paid to maintaining good husbandry conditions on farms. Animal welfare qualities have been known for many years [1,2]. This term has been connected to many issues that are mainly related to the provision of a proper animal physical and mental state. Nowadays, there is a great deal of controversy over performing invasive procedures on farm animals, such as castration, teeth clipping, and tail-docking—especially when performed without anesthesia [3,4]. The pressure from consumers and animal welfare organizations regarding these and others aspects of animal rights is often reflected in legislation changes. At the same time, even the knowledge of professionals, official inspectors, and advisers in identifying the causes of tail biting and methods to reduce this behavior, as well as the understanding of legislation, is often insufficient [5].

One of the most important legal documents providing the minimum standards for pigs is the Directive 2008/120/EC [6]. This legislation states, among others, that pigs kept in groups should have access to litter or other materials that provide the possibility for exploration and occupation. Enrichment of the pigs' environment is a way to reduce aggression and other pathological behavior [7]. One of the greatest problems in the intensive production of pigs is tail biting. Taylor et al. [8] described the different motivations for performing this behavior. The factors that may influence the expression of tail biting are as follows: improper diet, an absence or delay of food provision, gastrointestinal discomfort, poor health, genotype, too large density, and unfavorable microclimate conditions. An important factor that may lead to a higher frequency of this behavior is also the lack of a substrate or object to manipulate in the pigs' surroundings [7,8].

In 2016, the Commission Recommendation (EU) 2016/336 on the application of the Council Directive 2008/120/EC laying down the minimum standards for the protection of pigs regarding the measures needed to reduce the need for tail-docking was released [9]. As the document states, this procedure should never be performed routinely. The Commission Recommendation provides useful information on estimating the risk of tail biting, and stipulates improving the conditions on the farm first, before invasive procedures become necessary. Taking into consideration the aspect of enriching the pigs' environment, this document—more precisely than the directive [6]—describes and characterizes the various categories of materials that may be used to improve animal welfare. The four-point evaluation of the enrichment materials contained in the Commission Recommendation states that they should be edible, chewable, investigable, and manipulable. Moreover, as the document states that enrichment materials should be provided in such a way that they are of sustainable attraction for pigs (e.g., regularly replenished), they should be accessible for oral manipulation and should be provided in sufficient quantity. They should also be clean. Only if the materials meet all of the above-mentioned criteria, may they be considered as an optimal. Suboptimal materials possess most of the above-mentioned characteristics, and they should be combined with other materials in pig housing. The last group of enrichments are materials of marginal interest. They cannot fulfill all of the animal's needs, therefore they should be used together with optimal and suboptimal materials.

In spite of the fact that straw is considered as the optimal material for pig housing improvement (according to the characteristics of different enrichment materials included in the recommendation), in slatted floor systems, it is not easy to introduce it, as it may lead to blocking the manure system [10]. However, enrichment devices commonly used in conventional farming, made of metal and plastic, may be assigned only as materials of marginal interest. Therefore, there is a great need to create and implement new materials and forms of environmental enrichment in pig housing. It also seems very important to analyze the results of the studies concerning the effect of various enrichment devices, and to check whether the materials used in the experiments meet the requirements contained in the Commission Recommendation.

The aim of this paper is to review and systematize the current state of knowledge on the topic of the impact of environmental enrichment on pig welfare. This review covers the issues of using the various materials to improve the housing of different pig groups, piglets pre- and post-weaning, and fattening pigs. In this review, the latest scientific literature has been included, mainly concerning the effect of enrichment materials on the reduction of undesirable behavior, such as tail biting, aggression, and stereotypies.

2. Welfare Problems in Pigs Resulting from Intensive Production Systems and the Methods of Welfare Assessment

2.1. The Concept of Animal Welfare and Assessment Methods

Animal welfare has been a scientific concept for many years [11]. It can be evaluated by scientific methods, and it may range from very poor to very good [12]. The term animal welfare covers, among others, the importance of the animal's ability to keep control of mental and body stability in different environmental conditions [13]. Broom [1] has established the definition of welfare as a state of an individual with regards to its attempts to cope with its environment. Animal welfare is poor when behavioral, physiological, immunological, and other brain coordinated component coping strategies fail. Indicators of welfare based on an animal's physiology and behavioral changes have been commonly used in recent years [12]. Among others, the presence of disease, injury, social interaction in the herd, housing conditions, and human handling are factors affecting animal welfare [12]. Improper social, hygienic, and microclimatic conditions may cause difficulties in maintaining the animal's mental and physical stability. Although an individual may adapt to unfavorable conditions, it may still suffer—feeling pain or frustration. Thus, adaptation does not always mean good welfare [12]. Currently, positive welfare indicators have been increasing in popularity [12,14,15]. Three of these behavioral indicators, as a tool for welfare assessment, were proposed by McCormick [15], as follows: the indicator of contentment/pleasure, the indicator of luxury behaviours, and the indicator of behaviours that support the ability to cope with challenge. Moreover, positive emotions in pigs may be indicated by play, barks, and tail movement assessment, while negative emotions are indicated by ear movements, tail low, freezing, defecating, urinating, escape attempts, and high-pitched vocalisation [16].

The occurrence of undesirable behavior may be considered as an indicator of poor welfare [13,17]. In assessing animal welfare, the scoring of injuries and skin damage is a very important tool [18,19]. Skin lesions noticed on the front body part of pigs reflect the high frequency of fighting among individuals in the pen. In turn, skin damage found on the middle region of the pig body does not show a significant relationship with the occurrence of this behavior. Skin damage on the rear part (including the tail) indicates that the individual has spent a large proportion of time being bullied.

Besides behavioral indicators, physiological measures are very commonly used to assess animal welfare. Physiological indicators of animal welfare, which are often used in animal studies, are measurements of stress hormones [20]. Improper environmental conditions evoke an animal body response, such as alterations in the hypothalamo–pituitary–adrenocortical (HPA) axis, whose consequence is, among others, the release of glucocorticoids [21]. When discussing animal welfare, it may be stated that a more important issue is the situation of chronic rather than acute stress. Chronic stress is a situation when the animal is unable to cope with its environment. It is worth adding that according to Munsterhjelm et al. [22], a barren environment compared with an enriched one is associated with signs of chronic stress. The long-term elevation of cortisol levels leads to impairment deterioration of the physical and emotional state [23,24]. The measurement of cortisol in blood (serum) has been a popular tool for the assessment of animal welfare, but currently, the search is on for methods that are less invasive than blood sampling [25,26]. Animal handling and immobilization may prove a threat, and can themselves constitute a source of anxiety for an animal and, therefore, induce the secretion of stress hormones [25]. Glucocorticoids measured in saliva, faecal, or hair samples are also reliable methods for the activation of the hypothalamic-pituitary-adrenal (HPA) axis, and the sampling procedure of these materials is less invasive [27,28]. However, taking into consideration that stress responses consist of different systems within the organism, the cortisol level is just one of the physiological parameters that should be evaluated [29]. The review of Mkwanazi et al. [30] raises many issues linked with various enrichments in pig housing. Among others, the performance and blood metabolites in pigs kept in enriched and barren pens were discussed by the above-mentioned authors. There has been some evidence that environmental enrichment may have a positive effect on growth rate and lower creatine kinase [31,32].

A wide range of behavioural methods used together with, preferably, non-invasive biomarkers, is of key importance for the reliable assessment of animal welfare.

2.2. The Natural Behavior and Needs of Pigs

When discussing animal well-being concepts, it must also be stated that animals have strong motivational systems to meet their essential needs. These include basic needs for eating and drinking, but also for exploring the surroundings and expressing their natural behavior [12,33]. Rooting is one of the most important activities for pigs; it is performed to forage, sows root to build a nest, and rooting may also have some thermoregulatory functions [34,35]. It seems that the absence of materials for rooting, foraging, or manipulating causes the direction of the pig's attention to be directed to other individuals in the pen [36,37]. According to some authors, the majority of pigs in Europe are reared in conditions of intensive production, in a barren environment with slatted floors [38,39]. This type of housing does not favor the expression of species-specific behavior. It has been shown that a lack of possibilities for being able to explore the surroundings in intensive production may lead to increased incidents of aggression, cannibalism, tail biting, and stereotypies [40–42].

Play may be considered as a luxury behavior, and thus as a positive indicator of welfare [15,16]. It is very important, especially for young pigs. Providing objects to play with in the piglets' environment may improve their social skills. It has been suggested that animals reared in an environment that enables expression of play behavior are better prepared to cope with unfavorable situations at a later stage of life [43].

2.3. Risk Factors for Undesirable Behaviours under Intensive Conditions

Tail biting is a huge problem for both the economic and welfare aspects of pig production. Taylor et al. [8] identified three different forms of tail biting, namely: two-stage, sudden-forceful, and obsessive. Two-stage tail biting is a situation when pigs gently manipulate another individual's tail, and at some point in this "routine", the skin damage occurs. Fresh blood may attract other pigs to bite, and it consequently, may lead to increased aggression within the herd, and even cannibalism in some cases [44]. This form of tail biting may be prevented by providing some materials to develop a manipulatory interest. Another form of this behavior may be characterized as an acute and rapid attack performed without any previously gentle manipulation of the tail. It is often seen when pigs have to compete for access to water or food [45]. The obsessive tail biting is mainly represented by one or a few individuals. Tail biters seem to be focused on this body part, and continuously look for another one to grab or bite. The occurrence of this pathology may be linked with unsuitable feed and genetic lines, and obsessive tail biting may be seen as some form of stereotypies.

An increase in aggressive behavior is often observed during mixing, as this is generally considered a stressful procedure, when the individuals of different herds are placed together in one pen. The social hierarchy must be established, so it is often linked with fighting and aggression [46]. Only after the herd hierarchy is established, the fights among the individuals of the pen gradually decline [47,48].

Another form of behavior observed mainly in weaned piglets, especially in those weaned earlier, is the belly nosing phenomenon characterized by the rhythmical rubbing of a piglet's snout on the other individual's belly [49]. Widowski et al. [50] described in detail the various aspects related to the occurrence of this stereotypic behavior. One of the reasons for belly nosing or belly suckling is early separation from the sow. The performing of this behavior among piglets for a longer period of time often leads to lesions, and also may negatively impact the pigs' performance [50]. These authors presented some evidence about the complex conjunction of belly nosing, drinking, feeding, and suckling in weaned piglets. Briefly, the strong motivation to suckle, together with an insufficiently developed ability to independently feed in early weaned piglets may be a reason for oral behavioral problems. As Breuer et al. [51] claimed, belly nosing may also have a genetic basis or may be a general behavioral reaction to stressful situations [52]. An environment with poor stimuli may also contribute to the development of this behavior [53,54].

There has been some evidence of how stereotypic behavior is induced in farm, laboratory, and wild animals kept in a barren environment [55–57]. Stereotypies are defined as a simple ritualized, non-functional, repetitive behaviors [58]. Situations in which the animal has a strong motivation to meet its needs, but cannot reach them, are frustrating. The results of this frustration may be reflected in substitute behavior or stereotypies [59]. Moreover, some stereotypies may be an attempt to reduce the physiological symptoms of stress. This behavior may lead to reduced anxiety and the ability to react to external harmful stimuli, as the animal's attention is diverted from the source of the conflict [59]. Stereotypies is mainly linked with the functioning of the basal ganglia. Briefly, it may be assumed that the behaviors associated with the continuous repetition of activities have their source, among others, in the disturbed balance between the activation of the two main basal ganglia regulatory loops [60].

2.4. Environmental Enrichment as an Effective Prevention/Mitigation Strategy for Tail Biting

In animal models, it has also been shown that environmental enrichment for the prevention of stereotypies is linked with a higher level of neuronal metabolic activity and dendritic morphology (more dendritic spines) in the motor areas of the brain [61]. Taking into account the multi-factorial nature of tail-biting risk, the provision of good enrichment is an important starting point, but it may not be enough to prevent an outbreak of tail biting [62]. When assessing the level of well-being, it should be observed how pigs are involved in interacting with their environment. It is necessary to check whether pigs are able to express appropriate exploration/manipulation behaviors; whether the enrichment material is edible, chewable, and destructible; and whether pigs deal with playpen elements instead of enrichment objects/materials. According to the directive, it is also important to find out whether all pigs have permanent access to a sufficient amount of enrichment materials that is safe and clean (EUWelNet training tool). This evaluation may constitute the basis for the selection of the best environmental enrichment solutions.

3. Different Types of Enrichment

The first idea for enriching the environment was proposed for laboratory animals in the 1940s [63]. Since that time, many studies have been carried out regarding the positive effect of an improvement in the animal environment on its brain structure and biochemical changes [64]. Environmental enrichment should stimulate animals' visual, somatosensory, and olfactory systems, and the key idea is that these objects should provide an aspect of novelty [65].

Taking into consideration the fact that pigs may lose attention towards the object within a few days [66], it is of key importance to sustain the animals' interest by the frequent replacement or renewal of enrichments. Moreover, the pigs should have enough materials to play with at the same time. These explanatory notes were included in the aforementioned Commission Recommendation (EU) 2016/336 [9]. Based on different categories of enrichment, straw, but also other materials such as green fodder, miscanthus pressed or chopped, and root vegetables, when used as bedding, may be considered as optimal for animal welfare [9]. As van de Weerd et al. [67] claimed, fragrance, chewability, and deformability are features of objects that, at the beginning attract the pigs, however, the destructibility and edibility of the enrichment may attract them for a longer time. In the Commission Recommendation, peanut shells, fresh wood, corn cobs, natural ropes, compressed straw cylinders, shredded paper, pellets, and so on are considered as suboptimal. Moreover, straw provided in racks or in dispensers is treated as a suboptimal material as well [9]. The marginal materials include, among others, chains, rubber, soft plastic, pipes, hard woods, and balls. For the best results when bedding cannot be provided, the combination of different kinds of enrichment materials should be used [9].

It seems that in spite of the fact that many years have passed since the Council Directive 2008/120/EU was introduced, there are still some experiments carried out using objects and materials not admitted by this legal EU Act, showing some positive effect on pigs' behavior. However, the earlier results obtained by Scott et al. [68] and Zwicker [69] indicate that straw bedding much more effectively ensures a high interest in this type of enrichment, compared with hanging toy(s), even when the number of hanging objects is increased. Moreover, straw provided in racks increases exploratory behaviour, especially when more enrichment of this type is provided. Hanging elements do not provide as much interest for pigs as straw, because they are not rootable, and most of them are not edible, so they do not fulfil the requirements from the recommendation. It is worth adding that the Recommendation, in comparison to the directive, contains many more valuable details.

The review of the literature included in the report of Ernst et al. [66] covers the majority of different solutions used for environmental enrichment in pigs' conventional production. Some authors confirm the benefits of using different kinds of enrichments at the same time [66,70]. Before the Commission Recommendation 336/2016 was introduced into practice, the objects most often used were metal chains or commercially available plastic or rubber toys [71]. There is some evidence that this may still be the case. For example, the final report of an audit carried out in Germany from 12 February 2018 to 21 February 2018 in order to evaluate member state activities to prevent tail-biting and avoid routine tail-docking of pigs [72] showed that the German authorities can impose administrative fines for non-compliances with directives. But it was noted that some requirements of the provisions of the Council Directive 2008/120/EC are not directly sanctionable through this system, in the case of the characteristics and sufficient quantities of the enrichment material. Moreover, although central and land authorities have spent considerable sums on research and communicating its results, their strategies to reduce tail biting and avoid the routine tail-docking of pigs have not produced tangible results, and tail-docking is still routinely carried out in Germany. The central competent authority estimates the incidence of tail-docking in Germany is over 95%. In addition, the French Agency for Food, Environmental, and Occupational Health and Safety consider chains "acceptable" as an enrichment [73]. In addition to the optimal materials and objects listed in the Commission Recommendation, environmental enrichment may also be a larger space, fragrance, or even music [57,74,75]. Nutrition can be a special enrichment, especially the way it is provided to animals (e.g., hidden in the substrate) may have some key benefits for improving the pigs' activity [16]. Nowadays, various forms of cognitive enrichment, linked mainly with a reward gained by an animal, may bring many advantages regarding pig welfare [76]. Reimert et al. [16] found that rewarding events both during training and testing in pigs are often linked with an increase in play behavior. The authors observed also that barking and tail wagging occurrence is more common in pigs during rewarding events. Thus, it may generally be considered that a reward gained by an animal is associated with eliciting positive emotions. However, it should always be remembered that the legislation requirements should be fulfilled.

4. Environmental Enrichment for Sucker and Weaned Piglets

The effects of environmental improvement have been studied in pre- and post-weaning piglets. It seems that providing the enrichment in early, neonatal housing conditions may lead to better social behavior at later stages of the pig's life [75,77,78]. As has been shown, access to straw, wood shavings, wood bark, or even pieces of newspaper may have a positive effect on a piglets' behavior at the pre-weaning stage [79–81].

The effect of straw provision at an early stage may be effective in the reduction of nosing and tail biting [77]. Access to straw at the pre-weaning period may also lead to a reduction in mounting and oral manipulation directed at other individuals, as well as positively affect piglets' growth [80,82]. Moreover, Brajon et al. [82] observed more time spent lying in piglets that had access to straw in the pre-weaning period, which may be considered as a positive indicator of animal comfort.

Positive indicators of animal welfare may also be observed when materials other than straw are provided. As mentioned previously, belly nosing has a multifactor origin, and generally may be considered as an indicator of poor welfare in weaned piglets [82]. Previously, it was documented that straw provision may not be effective in the total elimination of this phenomenon [50,82,83]. However, some positive aspects of belly nosing reduction were found when black foam rubber matting was used [84]. Similarly, in other studies, the most effective influence on some of the behavioral and physiological indicators were materials classified according to the recommendation as suboptimal or of marginal interest [79,81]. Yang et al. [79] found that play behavior was performed more often by piglets of the two groups that had access to enrichment (hanging objects or wood bark) than individuals kept in farrowing pens without any improvement. In turn, in the study of Telkänranta et al. [81], all of the experimental animals were housed in pens with environmental enrichment. However, the individuals of the control group only had access to balls and wood shavings, and the piglets of the experimental group had access to sisal ropes, plastic balls, pieces of newspaper, and wood shavings. The sisal ropes and newspaper pieces in particular induced the pigs' activity, as well as influenced a reduction of oral–nasal manipulation directed towards pen mates.

Further effects of environmental enrichment provided in the neonatal period were shown in some studies [75,80–83].

It is well known that chronic stress has an immunosuppressive effect [85]. Taking this aspect into consideration, it worth referring to the studies of Yang et al. [79] and Van Dixhoorn et al. [80]. The first mentioned authors found that piglets that had access to wood bark at the neonatal stage had a lower level of cortisol after weaning. In turn, the latter authors found that piglets kept in a pen with straw, moist peat, and wood shavings at the pre-weaning stage were characterized by a less severe onset of the reproductive and respiratory virus and *Actinobacilllus pleuropneumoniae* co-infection at a later stage of their lives.

In turn, Telkänranta et al. [81] reported that richer environmental enrichment provided in the neonatal period caused reduced severe tail damage (wounds with swelling and infection or partial/total loss of the tail). This effect was not seen in the Yang et al. [79] study. No significant impact on aggressive behavior was found; however, the aggression level was evaluated through the skin lesion pre-weaning, and two days after mixing. However, perhaps more days of observation were needed. Martin et al. [75] compared the behavior of piglets kept in standard neonatal housing conditions and in improved (larger and with straw provision) pens. These authors claimed that three days post-weaning, the aggression was similar, regardless of whether the enrichment had been provided or not, but after seven days, a significant drop in this behavior was observed in the enriched group.

Brajon et al. [82] found that the individuals that were stimulated during the neonatal period performed less nosing and chewing behavior towards pen mates than the piglets of the control group, six days after weaning. The situation changed after six days; belly nosing became significantly higher in this group and activity levels decreased. This was probably the effect of stress, linked with weaning and enrichment removal.

As has been well documented in previous studies, the sudden limitation of enrichments may cause an increase in harmful social behavior [86,87]. This effect was noted in the Statham et al. [83] study. The authors carried out a longitudinal experiment, whose aim was to evaluate the impact of straw provided through the entire production cycle. The behavior of the animals assigned to four groups was observed. The first group was housed with straw through the entire production cycle (from farrowing to finishing pens); in the second group, straw was provided from weaning; and the third group was kept on straw only in the finishing pen. The fourth group had no straw provided; however, prior to weaning, the pens of the third and fourth groups had a small amount of wood shavings added. Taking into consideration the proportion of the diverse behavior categories performed by piglets at the pre-weaning stage, there was no significant difference between the treatments. Moreover, the authors claimed that access to straw throughout the animal's life did not affect the level of aggression, belly nosing, or tail biting. This lack of differences between treatments may result from the pens' construction, which enabled the adding of a greater amount of straw in the early stages than at the finishing stage. This could induce the aforementioned negative effect of enrichment removal in animals. Moreover, the lack of significant differences between treatments in this study could also be related to the insufficient frequency of straw provision (twice a week); novelty in the pigs' surroundings may be ensured when straw is added more often [78]. In addition, it is worth mentioning that all of the piglets had access to enrichment (straw or wood shavings) at weaning, so possibly, the effect of both of these substrates was comparable.

Aromas in pigs' surroundings may be a factor causing positive stimulation, which has been proven in some studies [84,88,89]. In the Bench and Gonyou [84] experiment, a soil-filled tray was very effective in the improvement of the weaned pigs' manipulating behavior. The pigs were probably attracted by the smell of this enrichment. Nowicki et al. [88] found that among the synthetic and natural aromas, the weaned pigs preferred the most natural fragrances, for example moist soil, fresh grass, and dried mushrooms. Moreover, at the second stage of their experiment, the authors compared the behavior of pigs kept in three different pens. The first pen was without enrichment, the second had an aromatized object (firstly with moist soil, then with fresh grass), and in the third pen was placed a container without any aromas. The aromatized containers had the greatest impact on reducing the time spent on agonistic behavior during the first nine days of observation. After this time, there was no significant difference between the treatments. It was also stated that the possibility of aroma exchange after several days of use provides the required novelty, and makes the enrichment attractive again. In turn, Sartor et al. [89], in addition to different fragrances, used Vivaldi's music and blue light emitting diode—LED lightening as enrichments in their study. The experiment was performed over 10 days on sucker piglets. Of the fragrances of chamomile, lavender, lemon, and thyme used in the heated creeps, the piglets preferred the latter aroma the most. Blue lightening was also an important factor that caused longer time spent in the creep by the piglets. The classical music stimuli, however, affected higher activity among the piglets.

Other environmental enrichment than that mentioned previously was tested by Winfield et al. [90]. The authors evaluated different shapes of nutritional lick blocks on sucker and weaned piglets' behavior. Taking into consideration the wedge-, brick-, and cube-shaped blocks, the authors found that the brick shape attracted the piglets the most. Multiple piglets would stand together to chew the block. This behavior was similar to co-operatively massaging a sow udder, thus the authors suggest that this shape of enrichment may be a factor evoking natural co-operative behavior.

Weaned piglets experience multiple stressors while they are subjected to nutritional and social changes [91]. The literature references presented in this section show some evidence that the environmental stimuli may alleviate the negative effects of this treatment. Even the presence of suboptimal materials in the pen after weaning was shown to be effective at reducing harmful behaviour during the first period after mixing [88]. A key aspect also seems to be to provide the enrichment at the neonatal stage.

Table 1 shows some current literature references regarding enrichment for sucker and weaned piglets, and includes the main effects of each type of enrichment on pig behavior. It also shows the categories of environmental enrichment in reference to the Commission Recommendation.

Table 1. Environmental enrichment materials and objects according to the categories included in the Commission Recommendation (EU) 2016/336, and their main effects on sucker and weaned piglets' behavior.

Animal	Enrichment	Duration of Behaviour Observation	Main Effect	References
sucker/weaned piglets	black foam rubber matting ***	8 h daily on 3, 10, 19, 26, 33 d of pig's life	reduction in belly nosing	Bench and Gonyou [84]
sucker/weaned piglets	farrowing pen—straw bedding *	6 h daily on 6, 12, 20, 21, 22, 27 d of pig's life	increased level of exploration, animal comfort improved	Brajon et al. [82]
sucker/weaned piglets	farrowing pen—wood shavings **, sisal rope **, plastic ball attached to chain ***, newspaper **	4 h daily one day per 2, 3, 9 wks. of pig's life	at pre-weaning stage—decrease in pen mate manipulation	Telkänranta et al. [81]
	weaners pen—wood shavings **, sisal rope **, plastic chewing toy ***		at post-weaning stage—decrease in severe tail damage	
sucker/weaned piglets	farrowing pen—wood bark in a plastic box **	6 h daily on 4, 5, 11, 12, 19, 23 d of pig's life	baseline of cortisol level on d 1 post-weaning	Yang et al. [79]
weaned piglets	aromatized plastic container **	24 h per 9 d after weaning	decrease of duration of agonistic behavior	Nowicki et al. [88]

* optimal, ** suboptimal, *** marginal materials according to the categories included in the Commission Recommendation (EU) 2016/336 on the application of the Council Directive 2008/120/EC, laying down the minimum standards for the protection of pigs with regards to measures to reduce the need for tail-docking [9].

5. Environmental Enrichment for Fattening Pigs

Taking into consideration the environmental enrichment of growing and finishing pigs, a larger space seems to be a key factor for providing animal comfort. In the study of Cornale et al. [39], the faecal corticosteroids level was found to be higher in the pigs kept in pens with a greater number of individuals, even when access to enrichment (cylindrical pieces of hard wood suspended on a chain) was provided. In turn, the experiment of Di Martino et al. [92] showed that at high stocking density conditions, the cases of tail lesions were high, even when the animals had access to straw racks. These results may suggest that a larger space is a very important factor that may reduce this pathology.

It is worth adding that the fattening heavy pigs used for specific productions require the provision of more space than the minimum legal requirements. These situations may be seen when pigs reach 180–190 kg of body weight, and when the fattening process is long [92]. The Council Directive 2008/120/EC states that pigs should benefit from an environment corresponding to their needs for exercise and investigatory behavior. The welfare of pigs appears to be compromised by severe restrictions of space. The 1 m^2 space for pigs heavier than 150 kg seems to be insufficient. Moreover, the directive states that before carrying out tail docking, other measures should be taken in order to prevent tail-biting and other vices, taking into account environment and stocking densities. For this reason, inadequate environmental conditions or management systems must be changed.

The genetic background is also a significant factor of tail biting occurrence [7]. This was confirmed in the study of Bulens et al. [93]. The authors compared the effect of two methods of enrichment—straw block and chain versus only chain provision—for finishing pigs of two genetic backgrounds. The boar type was the factor for determining the frequencies of tail biting, regardless of the enrichment provided.

Some positive aspects of the prevalence of mild damage to the tail and ear biting reduction were found in the Telkänranta et al. [70] study, by providing pieces of freshly cut birch trees. This enrichment was the most effective in the mentioned aspects, but the authors also tested different materials. The pens of the control group were provided with a straw dispenser, wood shavings, and metal chain; the four other groups had the same enrichments and some other materials such as a horizontally suspended piece of fresh birch wood, polythene pipes, and metal chains suspended vertically. The last group was housed in the pens equipped with all of the enrichment objects mentioned above. The other result of this study was the finding that freshly cut birch wood and polythene pipe, among other enrichment devices, significantly affected the frequency of object manipulation.

Continued high interest in straw investigation was found in the study of Scott et al. [68], when this substrate was used as a bedding. However, the study of Nannoni et al. [94] confirmed that even when bedding is not provided, finishing pigs in all production cycles were interested in exploring the environmental enrichment inside the racks. These authors tested wooden logs and a vegetal edible block placed inside the metal racks compared to hanging chains. Taking into consideration the effect of enrichment materials on tail biting cases, the statistical analysis showed significant differences only between the moderate tail lesion noticed in the group with access to the wooden logs, compared to the group with the chain. Despite the authors' suggestion that the vegetal edible block helped to reduce the exploratory behavior directed towards the pen floor, it seems it could not fully reduce the pigs' attention directed towards other pigs. The group that had access to this enrichment was characterized with some moderate and severe tail lesions. This could be a result of competition for access to enrichment, as was also shown in the study of Bulens et al. [93]. In this experiment, mounting and fighting was higher at the beginning of the finishing phase in the pigs that had access to the straw dispenser.

The effect of the enrichment on stress response in finishing pigs was tested in Casal et al.'s [28] and Nannoni et al.'s [94] research. In the Casal et al. [28] study, the pens of the experimental groups contained sawdust, natural hemp ropes, and rubber balls, whereas the control group was kept in a barren environment. The authors also tested herbal supplementation. After two months of treatments, the hair cortisol levels and salivary chromogranin A (CgA) were the highest in the control group. Consequently, the authors suggest that both enrichment and herbal compounds may lead to stress reduction in pigs. However, the enrichment devices used in Nannoni et al.'s [94] research had no impact on the hair cortisol levels.

The above-mentioned findings [28,93,94] were obtained through a long-term study. This way of performing an experiment seems to be very important for the reliable assessment of the impact of different categories of enrichment, especially when the markers of stress are used. However, the impact on the behavior of the environmental improvement was found during only six days of observation in the Machado et al. [95] study. One of the trials performed in this research was to evaluate the effect of the availability of hanging toys (made from a polyvinyl chloride—PCV pipes and plastic tubing). The pigs of the control group had no access to the enrichment, the individuals of the first group had the enrichment for the experimental period, the pigs of the second group had the objects in their pens but provided on alternate days, and the pigs of the last group were provided with the objects in the morning and then had them removed in the afternoon. The results showed that the individuals in the pen without enrichment and those that had access to the objects for the entire time spent more time eating and drinking. However, as suggested by the authors, the individuals of the pen without objects might have wasted the food during play. The higher percentage of sexual behavior (mounting) was found among the pigs with access to the enrichment provided on alternate days. It could be the same effect as seen in previously presented studies [93,94]; when access to enrichment is limited, competition rises, and consequently, a negative form of behaviour may occur. In the Machado et al. [95] study, resting behavior was performed the most frequently in the group with no object, compared with individuals that had access to enrichment constantly or on alternate days. There were no differences, however, between the treatments regarding the time spent on interacting with other pigs, nuzzling, or exploring the environment. Moreover, the objects had no impact on agnostic behavior, which generally was quite low during the experimental period.

One of the main findings of this study was the increase of animal activity, even when only the material classified by the EU Recommendation as being of marginal interest is provided. The animals of the control group which had no access to a "toy" remained for the majority of time inactive. Based on this and other results presented in this section, regardless of fact that harmful behavior still occurred in the pigs kept with access to enrichment, it may be stated that the increase of animal activity and interest in an enrichment object is very valuable from an animal welfare perspective. Moreover, Scott et al. [68] pointed out the greater benefits of using straw as an environmental improvement to hanging toys; however, in the Telkänranta et al. [70] study, it was proven that fresh wood attracted pigs the most, even when, at the same time, they had access to straw. These kinds of results are promising for pigs housing, where the technological system of maintenance makes straw provision limited or simply impossible. On the other hand, it should be stated here that in practice within EU countries, enrichment materials classified as being of marginal interest cannot be used alone (without provision of optimal enrichment).

Similarly, as in Table 1, Table 2 shows some literature references regarding enrichment, but for fatteners, including the main effects of each type of enrichment on the behavior of pigs. It also shows the categories of environmental enrichments in reference to the Commission Recommendation.

Table 2. Environmental enrichment materials and objects according to the categories included in the Commission Recommendation (EU) 2016/336, and their main effects on fattening pig behavior.

Animal	Enrichment	Duration of Behavior Observation	Main Effect	References
fattening pigs	straw bedding * compared to slatted floor pens with different number of hanging toys ***	6 h daily in the week of entry, the week before the group size reduction and a week before slaughter	decrease of pen mate manipulation and increase of investigation of straw bedding	Scott et al. [68]
fattening pigs	straw rack**, metal chain***, wood shavings**, and additional pieces of fresh birch wood**, polythene pipes***, or two crosses of metal chain***; combination pens had all of the enrichment materials mentioned above	after 2.5 months of exposure to the objects, one day of video observation was performed	increased object exploration best seen in pens with fresh wood, as well as reduce of tail and ear biting	Telkänranta et al. [70]
fattening pigs	piece of hard wood attached to chain **	3 h daily, nine times over 16 weeks	decrease of aggression and tail biting	Cornale et al. [28]
fattening pigs	chain *** and compressed straw block in dispenser *	2.5 h daily, once a week through the entire finishing stage	decrease of pen mate manipulation	Bulens et al. [93]
fattening pigs	PCV pipes with four pieces of plastic tubing attached to it ***	8 h daily, for six consecutive days	increase of activity	Machado et al. [95]

* optimal, ** suboptimal, *** marginal materials according to the categories included in the Commission Recommendation (EU) 2016/336 on the application of the Council Directive 2008/120/EC, laying down the minimum standards for the protection of pigs with regards to measures to reduce the need for tail-docking [9].

6. Conclusions

Based on the various research results presented in the review, it may be stated that the enrichments provided for young and fattening pigs have many benefits with regard to animal welfare. Generally, it may be concluded that access to even a small amount of substrate and objects increases targeted activity. As it has currently been proven, enrichment given at the neonatal stage (farrowing pen) improves manipulating skills, and may also lead to better social behavior performed at later stages of the animal's life. The main effects of enrichment may be seen through a reduction of attention paid to other individuals in the pen. This also includes the pigs' interest in performing tail chewing, biting, or belly nosing. However, the results of a reduction in these harmful behaviors through the provision of enrichments is inconclusive. Despite the fact that studies have confirmed the advantages of using different kinds of enrichments at the same time, some results have also shown that only the use of materials that are considered to be of marginal interest (according to Commission Recommendation (EU) 2016/336) brought about the desired effect on the pigs' behavior. It can be misleading for farmers, because it is strongly highlighted in the recommendation that "materials of marginal interest—materials providing a distraction for pigs, which should not be considered as fulfilling their essential needs, and therefore optimal or suboptimal materials should also be provided". Moreover, some authors found a positive impact of materials, such as newspaper or PCV pipes, with protruding tubes. As such, it remains an open question whether these enrichments could be considered as completely safe for animal health; similarly for hard wood, whose chewing may also result in injury.

It is not possible and it makes no sense to change the European legislation regulating the types of environmental enrichment materials and their features. The current EU regulations are based on research results and represent the results of the European Union scientific projects. It should always be taken into account that tail biting and ear biting have a multifactorial origin, which is often difficult to determine. Tail-docking should be an absolute last resort (in some of the EU countries it is forbidden), so looking for alternative possible ways in terms of scientific and practical activities to stop and prevent tail biting should be essential (but always in agreement with the legal regulations). Making changes in the pigs' environment, such as reducing stocking density and introducing optimal environmental enrichment solutions based on the scientific evaluation of their efficiency, should lead to the situation in which the procedure of tail-docking would be performed as little as possible, or not at all.

Author Contributions: Conceptualization and writing (original draft preparation and supervision), D.G.; writing (review and editing), J.N. and P.H.

Funding: This research received no external funding.

Acknowledgments: The authors would like to thank Natasza Dobrowolska for her assistance and engagement.

Conflicts of Interest: The authors declare no conflict of interest.

References

1. Broom, D.M. Indicators of poor welfare. *Br. Vet. J.* **1986**, *142*, 524–526. [CrossRef] [PubMed]
2. Fraser, D.; Weary, D.M.; Pajor, E.A.; Milligan, B.N. A Scientific Conception of Animal Welfare that Reflects Ethical Concerns. *Anim. Welf.* **1997**, *6*, 187–205.
3. Godyń, D.; Herbut, E.; Walczak, J. Infrared thermography as a method for evaluating the welfare of animals subjected to invasive procedures—A Review. *Ann. Anim. Sci.* **2013**, *13*, 423–434. [CrossRef]
4. Sutherland, M.A.; Davis, B.L.; McGlone, J.J. The effect of local or general anesthesia on the physiology and behavior of tail docked pigs. *Animal* **2011**, *5*, 1237–1246. [CrossRef] [PubMed]
5. Hothersal, B.; Whistance, L.; Zedlacher, Z.; Algers, B.; Andersson, E.; Bracke, M.; Courboulay, V.; Ferrari, P.; Leeb, C.; Mullan, S.; et al. Standardising the assessment of environmental enrichment and tail-docking legal requirements for finishing pigs in Europe. *Anim. Welf.* **2016**, *25*, 499–509. [CrossRef]
6. EC Directive 2008/120/EC. Available online: https://eur-lex.europa.eu/legal-content/EN/TXT/PDF/?uri=CELEX:32008L0120&from=EN (accessed on 16 April 2019).

7. EFSA—Panel on Animal Health and Welfare. The risks associated with tail biting in pigs and possible means to reduce the need for tail docking considering the different housing and husbandry systems. *EFSA J.* **2007**, *5*, 611. [CrossRef]
8. Taylor, N.R.; Main, D.C.J.; Mendl, M.; Edwards, S.A. Tail-biting: A new perspective. *Vet. J.* **2010**, *186*, 137–147. [CrossRef] [PubMed]
9. The European Commission. Commission Recommendation (EU) 2016/336 of 8 March 2016 on the application of Council Directive 2008/120/EC laying down minimum standards for the protection of pigs as regards measures to reduce the need for tail-docking. *Off. J. Eur. Union* **2016**. Available online: https://eur-lex.europa.eu/eli/reco/2016/336/oj (accessed on 16 April 2019).
10. Wallgren, T.; Westin, R.; Gunnarsson, S. A survey of straw use and tail biting in Swedish pig farms rearing undocked pigs. *Acta Vet. Scand.* **2016**, *58*, 84. [CrossRef]
11. Fraser, D. Understanding animal welfare. *Acta Vet. Scand.* **2008**, *50*, 1–7. [CrossRef]
12. Broom, D.M. A history of animal welfare science. *Acta Biotheor.* **2011**, *59*, 121–137. [CrossRef] [PubMed]
13. Broom, D.M. The scientific assessment of animal welfare. *Appl. Anim. Behav. Sci.* **1988**, *20*, 5–19. [CrossRef]
14. Boissy, A.; Manteuffel, G.; Jensen, M.B.; Moe, R.O.; Spruijt, B.; Keeling, L.J.; Winckler, C.; Forkman, B.; Dimitrov, I.; Langbein, J.; et al. Assessment of positive emotions in animals to improve their welfare. *Physiol. Behav.* **2007**, *92*, 375–397. [CrossRef] [PubMed]
15. McCormick, W. Recognising and Assessing Positive Welfare: Developing Positive Indicators for Use in Welfare Assessment. In Proceedings of the Measuring Behavior, Utrecht, The Netherlands, 28–31 August 2012; Spink, A.J., Grieco, F., Krips, O.E., Loijens, L.W.S., Noldus, L.P.J.J., Zimmerman, P.H., Eds.; pp. 241–243. [CrossRef]
16. Reimert, I.; Bolhuis, J.E.; Kemp, B.; Rodenburg, T.B. Indicators of positive and negative emotions and emotional contagion in pigs. *Physiol. Behav.* **2013**, *109*, 42–50. [CrossRef] [PubMed]
17. Mason, G.; Rushen, J. Stereotypic behaviour in captive animals: Fundamentals and implications for welfare and beyond. In *Stereotypic Animal Behaviour: Fundamentals and Applications to Welfare*, 2nd ed.; CABI: Cambridge, UK, 2006; pp. 326–356.
18. Turner, S.P.; White, I.M.S.; Brotherstone, S.; Farnworth, M.J.; Knap, P.W.; Penny, P.; Mendl, M.; Lawrence, A.B. Heritability of post-mixing aggressiveness in grower-stage pigs and its relationship with production traits. *Anim. Sci.* **2006**, *82*, 615–620. [CrossRef]
19. EFSA Panel on Animal Health and Welfare. Statement on the use of animal-based measures to assess the welfare of animals. *EFSA J.* **2012**, *10*, 2767. [CrossRef]
20. Mormède, P.; Andanson, S.; Auperin, B.; Beerda, B.; Guemene, D.; Malmkvist, J.; Manteca, X.; Manteuffel, G.; Prunet, P.; Van Reenen, C.G.; et al. Exploration of the hypothalamic-pituitary-adrenal function as a tool to evaluate animal welfare. *Physiol. Behav.* **2007**, *92*, 317–339. [CrossRef]
21. Morris, M.C.; Compas, B.E.; Garber, J. Relations among Posttraumatic Stress Disorder, Comorbid Major Depression, and HPA Function: A Systematic Review and Meta-Analysis. *Clin. Psychol. Rev.* **2012**, *32*, 301–315. [CrossRef]
22. Munsterhjelm, C.; Valros, A.; Heinonen, M.; Hälli, O.; Siljander-Rasi, H.; Peltoniemi, O.A.T. Environmental enrichment in early life affects cortisol patterns in growing pigs. *Animal* **2010**, *4*, 242–249. [CrossRef]
23. Keay, J.M.; Singh, J.; Gaunt, M.C.; Kaur, T. Fecal glucocorticoids and their metabolites as indicators of stress in various mammalian species: A literature review. *J. Zoo Wildl. Med.* **2006**, *37*, 234–244. [CrossRef]
24. Broom, D.M.; Fraser, A.F. *Domestic Animal Behaviour and Welfare*, 4th ed.; CABI: Cambridge, UK, 2007; pp. 438–439.
25. Stewart, M.; Webster, J.; Schaefer, A.; Cook, N.; Scott, S. Infrared thermography as a non-invasive tool to study animal welfare. *Anim. Welf.* **2005**, *14*, 319–325.
26. Kersey, D.C.; Dehnhard, M. The use of noninvasive and minimally invasive methods in endocrinology for threatened mammalian species conservation. *Gen. Comp. Endocrinol.* **2014**, *203*, 296–306. [CrossRef] [PubMed]
27. Cook, N.J. Review: Minimally invasive sampling media and the measurement of corticosteroids as biomarkers of stress in animals. *Can. J. Anim. Sci.* **2012**, *92*, 227–259. [CrossRef]
28. Casal, N.; Manteca, X.; Escribano, D.; Cerón, J.J.; Fàbrega, E. Effect of environmental enrichment and herbal compound supplementation on physiological stress indicators (chromogranin A, cortisol and tumour necrosis factor-α) in growing pigs. *Animal* **2017**, *11*, 1228–1236. [CrossRef] [PubMed]

29. Blache, D.; Terlouw, C.; Maloney, S.K. Physiology. In *Animal Welfare*, 2nd ed.; Appleby, M.C., Mench, J.A., Olsson, I.A.S., Hughes, B.O., Eds.; CABI: Cambridge, UK, 2007; pp. 155–182.
30. Mkwanazi, M.V.; Ncobela, C.N.; Kanengoni, A.T.; Chimonyo, M. Effects of environmental enrichment on behaviour, physiology and performance of pigs—A review. *Asian Australas. J. Anim. Sci.* **2019**, *32*, 1–13. [CrossRef] [PubMed]
31. Beattie, V.E.; O' Connell, N.E.; Moss, B.W. Influence of environmental enrichment on the behaviour, performance and meat quality of domestic pig. *Livest. Prod. Sci.* **2000**, *65*, 71–79. [CrossRef]
32. Peeters, E.; Geers, R. Influence of provision of toys during transport on stress responses and meat quality of pigs. *Anim. Sci.* **2006**, *82*, 591–595. [CrossRef]
33. Kittawornat, A.; Zimmerman, J.J. Toward a better understanding of pig behavior and pig welfare. *Anim. Health Res. Rev.* **2010**, *12*, 25–32. [CrossRef]
34. Burne, T.H.J.; Murfitt, P.J.E.; Gilbert, C.L. Influence of environmental temperature on PGF2α-induced nest building in female pigs. *Appl. Anim. Behav. Sci.* **2001**, *71*, 293–304. [CrossRef]
35. Olsen, A.W. Behaviour of growing pigs kept in pens with outdoor runs I. Effect of access to roughage and shelter on oral activities. *Livest. Prod. Sci.* **2001**, *3*, 245–254. [CrossRef]
36. Kelly, H.R.C.; Bruce, J.M.; English, R.R.; Fowler, V.R.; Edwards, S.A. Behaviour of 3 week weaned pigs in straw-flow, deep straw and flat desk housing systems. *Appl. Anim. Behav. Sci.* **2000**, *68*, 269–280. [CrossRef]
37. Studnitz, M.; Jansen, K.H.; Jorgensen, E. The effect of nose rings on the exploratory behaviour of outdoor gilts exposed to different tests. *Appl. Anim. Behav. Sci.* **2003**, *84*, 41–57. [CrossRef]
38. Guy, J.H.; Meads, Z.A.; Shiel, R.S.; Edwards, S.A. The effect of combining different environmental enrichment materials on enrichment use by growing pigs. *Appl. Anim. Behav. Sci.* **2013**, *144*, 102–107. [CrossRef]
39. Cornale, P.; Macchi, E.; Miretti, S.; Renna, M.; Lussiana, C.; Perona, G.; Mimosi, A. Effects of stocking density and environmental enrichment on behavior and fecal corticosteroid levels of pigs under commercial farm conditions. *J. Vet. Behav.* **2015**, *10*, 569–571. [CrossRef]
40. Beattie, V.E.; Walker, N.; Sneddon, I.A. Effects of environmental enrichment on behaviour and productivity of growing pigs. *Anim. Welf.* **1995**, *4*, 207–220.
41. Cox, L.N.; Cooper, J.J. Observations on the pre-and post-weaning behaviour of piglets reared in commercial indoor and outdoor environments. *Anim. Sci.* **2001**, *72*, 75–86. [CrossRef]
42. Scott, K.; Chennells, D.J.; Campbell, F.M.; Hunt, B.; Armstrong, D.; Taylor, L.; Gill, B.P.; Edwards, S.A. The welfare of finishing pigs in two contrasting housing systems: Fully-slatted versus straw-bedded accommodation. *Livest. Sci.* **2006**, *103*, 104–115. [CrossRef]
43. Spînka, M.; Newberry, R.C.; Bekoff, M. Mammalian play: Training for the un-expected. *Q. Rev. Biol.* **2001**, *76*, 141–168. [CrossRef]
44. Schrøder-Petersen, D.L.; Simonsen, H.B. Tail biting in pigs. *Vet. J.* **2001**, *162*, 196–210. [CrossRef]
45. Morrison, R.S.; Johnston, L.J.; Hilbrands, A.M. A note on the effects of two versus one feeder locations on the feeding behaviour and growth performance of pigs in a deep-litter, large group housing system. *Appl. Anim. Behav. Sci.* **2007**, *107*, 157–161. [CrossRef]
46. McGlone, J.J. Influence of resources on pig aggression and dominance. *Behav. Processes.* **1986**, *12*, 135–144. [CrossRef]
47. Parratt, C.A.; Chapman, K.J.; Turner, C.; Jones, P.H.; Mendl, M.T.; Miller, B.G. The fighting behaviour of piglets mixed before and after weaning in the presence or absence of a sow. *Appl. Anim. Behav. Sci.* **2006**, *101*, 54–67. [CrossRef]
48. Langbein, J.; Puppe, B. Analyzing dominance relationships by sociometric methods—A plea for a more standardized and precise approach in farm animals. *Appl. Anim. Behav. Sci.* **2004**, *87*, 293–315. [CrossRef]
49. Fraser, D. Observations on the behavioural development of suckling and early-weaned piglets during the first six weeks after birth. *Anim. Behav.* **1978**, *26*, 22–30. [CrossRef]
50. Widowski, T.M.; Torrey, S.; Bench, C.J.; Gonyou, H.W. Development of ingestive behaviour and the relationship to belly nosing in early-weaned piglets. *Appl. Anim. Behav. Sci.* **2008**, *110*, 109–127. [CrossRef]
51. Breuer, K.; Sutcliffe, M.E.M.; Mercer, J.; Rance, K.; Beattie, V.; Sneddon, I.; Edwards, S. The effect of breed on the development of adverse social behaviours in pigs. *Appl. Anim. Behav. Sci.* **2003**, *84*, 59–74. [CrossRef]
52. Salak-Johnson, J.L.; Anderson, D.L.; McGlone, J.J. Differential dose effects of central CRF and effects of CRF astressin on pig behavior. *Physiol. Behav.* **2004**, *83*, 143–150. [CrossRef] [PubMed]
53. Cooper, J.J.; Cox, L.N.; Whitworth, C. Early environmental experience and transferable skills in the weaned piglet. *Anim. Welf. Potters Bar* **2001**, *10*, S238.

54. Bolhuis, J.E.; Schouten, W.G.P.; Schrama, J.W.; Wiegant, V.M. Behavioural development of pigs with different coping characteristics in barren and substrate-enriched housing conditions. *Appl. Anim. Behav. Sci.* **2005**, *93*, 213–228. [CrossRef]
55. Mallapur, A.; Cheelam, R. Environmental influences on stereotypy and the activity budget of Indian leopards (*Panthera pardus*) in foru zoos in southern India. *Zoo Biol.* **2002**, *21*, 585–595. [CrossRef]
56. Tilly, S.-L.C.; Dallaire, J.; Mason, G.J. Middle-aged mice with enrichment-resistant stereotypic behaviour show reduced motivation for enrichment. *Anim. Behav.* **2010**, *80*, 363–373. [CrossRef]
57. Silva, F.R.; Miranda, K.O.D.S.; Piedade, S.M.D.S.; Salgado, D.D.A. Effect of auditory enrichment (music) in pregnant sows welfare. *Eng. Agríc.* **2017**, *37*, 215–225. [CrossRef]
58. Cronin, G.M.; Wiepkema, P.R. An analysis of stereotyped behaviour in tethered sows. *Ann. Rech. Vet.* **1984**, *15*, 263–270. [CrossRef] [PubMed]
59. Mason, G. Stereotypies: A Critical Review. *Anim. Behav.* **1991**, *41*, 1015–1037. [CrossRef]
60. Langen, M.; Kas, M.H.; Staal, W.G.; van Engeland, H.; Durston, S. The neurobiology of repetitive behavior: Of mice. *Neurosci. Biobehav. Rev.* **2011**, *35*, 345–355. [CrossRef] [PubMed]
61. Turner, C.A.; Lewis, M.H. Environmental enrichment: Effects on stereotyped behavior and neurotrophin levels. *Physiol. Behav.* **2003**, *80*, 259–266. [CrossRef] [PubMed]
62. Arey, D.S.; Franklin, M.F. Effects of straw and unfamiliarity on fighting between newly mixed growing pigs. *Appl. Anim. Behav. Sci.* **1995**, *45*, 23–30. [CrossRef]
63. Hebb, D.O. The effects of early experience on problem solving at maturity. *Am. Psychol.* **1947**, *2*, 306–307.
64. Reynolds, S.; Lane, S.J.; Richards, L. Using animal models of enriched environments to inform research on sensory integration intervention for the rehabilitation of neurodevelopmental disorders. *J. Neurodev. Disord.* **2010**, *2*, 120–132. [CrossRef] [PubMed]
65. Nithianantharajah, J.; Hannan, A.J. Enriched environments, experience-dependent plasticity and disorders of the nervous system. *Nat. Rev. Neurosci.* **2006**, *7*, 697–709. [CrossRef]
66. Ernst, K.; Ekkelboom, M.; Kerssen, N.; Smeets, S.; Sun, Y.; Yin, X. Play behavior and environmental enrichment in pigs. *WUR* **2018**, 1–59. Available online: https://www.wur.nl/upload_mm/e/f/b/6af2e2db-430e-4771-8f7d-6f5b974eab5e_final%20report%20ACT%202060%20juli%202018%20op%20website%20.pdf (accessed on 4 June 2019).
67. Van de Weerd, H.A.; Docking, C.M.; Day, J.E.L.; Avery, P.J.; Edwards, S.A. A systematic approach towards developing environmental enrichment for pigs. *Appl. Anim. Behav. Sci.* **2003**, *84*, 101–118. [CrossRef]
68. Scott, K.; Taylor, L.; Gill, B.P.; Edwards, S.A. Influence of different types of environmental enrichment on the behaviour of finishing pigs in two different housing systems: 2. Ratio of pigs to enrichment. *Appl. Anim. Behav. Sci.* **2007**, *105*, 51–58. [CrossRef]
69. Zwicker, B.; Gygax, L.; Wechsler, B.; Weber, R. Influence of the accessibility of straw in racks on exploratory behaviour in finishing pigs. *Livest. Sci.* **2012**, *148*, 67–73. [CrossRef]
70. Telkänranta, H.; Bracke, M.B.M.; Valros, A. Fresh wood reduces tail and ear biting and increases exploratory behavior in finishing pigs. *Appl. Anim. Behav. Sci.* **2014**, *161*, 51–59. [CrossRef]
71. Bracke, M.B.M.; de Lauwere, C.C.D.C.; Wind, S.M.; Zonderland, J.J. Attitudes of Dutch Pig Farmers Towards Tail Biting and Tail Docking. *J. Agric. Environ. Ethics.* **2012**, *26*, 847–868. [CrossRef]
72. European Commission. Final Report of an Audit Carried Out in Germany from 12 February 2018 to 21 February 2018 in Order to Evaluate Member State Activities to Prevent Tail-Biting and Avoid Routine Tail-Docking of Pigs; Ref. Ares (2018)4437429-29/08/2018; DG (SANTE) 2018-6445. 2018. Available online: http://ec.europa.eu/food/audits-analysis/audit_reports/details.cfm?rep_inspection_ref=2018-6445 (accessed on 4 June 2019).
73. Laloy, F.; Huet, F.R. Tail Docking towards a National Action Plan of French Pork Production 2017. Available online: https://circabc.europa.eu/sd/a/362f9971-7f5c-4ece-93b2-560efc445e73/National%20Plan%20on%20Tail%20Docking%20(FR)_LALOY%20F%20%26%20HUET%20R_2017_EN.pdf (accessed on 4 June 2019).
74. Nowicki, J.; Klocek, C. The effect of aromatized environmental enrichment in pen on social relations and behavioural profile of newly mixed weaners. *Ann. Anim. Sci.* **2012**, *12*, 403–412. [CrossRef]
75. Martin, J.E.; Ison, S.H.; Baxter, E.M. The influence of neonatal environment on piglet play behaviour and post-weaning social and cognitive development. *Appl. Anim. Behav. Sci.* **2015**, *163*, 69–79. [CrossRef]
76. Ernst, K.; Puppe, B.; Schön, P.C.; Manteuffel, G. A complex automatic feeding system for pigs aimed to induce successful behavioural coping by cognitive adaptation. *Appl. Anim. Behav. Sci.* **2005**, *91*, 205–218. [CrossRef]

77. Day, J.E.L.; Burfoot, A.; Docking, C.M.; Whittaker, X.; Spoolder, H.A.M.; Edwards, S.A. The effects of prior experience of straw and the level of straw provision on the behaviour of growing pigs. *Appl. Anim. Behav. Sci.* **2002**, *76*, 189–202. [CrossRef]
78. Moinard, C.; Mendl, M.; Nicol, C.J.; Green, L.E. A case control study of on-farm risk factors for tail biting in pigs. *Appl. Anim. Behav. Sci.* **2003**, *81*, 333–335. [CrossRef]
79. Yang, C.H.; Ko, H.L.; Salazar, L.C.; Llonch, L.; Manteca, X.; Camerlink, I.; Llonch, P. Pre-weaning environmental enrichment increases piglets' object play behaviour on a large scale commercial pig farm. *Appl. Anim. Behav. Sci.* **2018**, *202*, 7–12. [CrossRef]
80. Van Dixhoorn, I.D.E.; Reimert, I.; Middelkoop, J.; Bolhuis, J.E.; Wisselink, H.J.; Groot Koerkamp, P.W.G.; Kemp, B.; Stockhofe-Zurwieden, N. Enriched Housing Reduces Disease Susceptibility to Co-Infection with Porcine Reproductive and Respiratory Virus (PRRSV) and *Actinobacillus pleuropneumoniae* (A. *pleuropneumoniae*) in Young Pigs. *PLoS ONE* **2016**, *11*, e0161832. [CrossRef] [PubMed]
81. Telkänranta, H.; Swan, K.; Hirvonen, H.; Valros, A. Chewable materials before weaning reduce tail biting in growing pigs. *Appl. Anim. Behav. Sci.* **2014**, *157*, 14–22. [CrossRef]
82. Brajon, S.; Ringgenberg, N.; Torrey, S.; Bergeron, R.; Devillers, N. Impact of prenatal stress and environmental enrichment prior to weaning on activity and social behaviour of piglets (*Sus scrofa*). *Appl. Anim. Behav. Sci.* **2017**, *197*, 15–23. [CrossRef]
83. Statham, P.; Green, L.; Mendl, M. A longitudinal study of the effects of providing straw at different stages of life on tail-biting and other behaviour in commercially housed pigs. *Appl. Anim. Behav. Sci.* **2011**, *134*, 100–108. [CrossRef]
84. Bench, C.J.; Gonyou, H.W. Effect of environmental enrichment at two stages of development on belly nosing in piglets weaned at fourteen days. *J. Anim. Sci.* **2006**, *84*, 3397–3403. [CrossRef]
85. Connor, T.J.; Brewer, C.; Kelly, J.P.; Harkin, A. Acute stress suppresses pro-inflammatory cytokines TNF-α and IL-1β independent of a catecholaminedriven increase in IL-10 production. *J. Neuroimmunol.* **2005**, *159*, 119–128. [CrossRef]
86. Van de Weerd, H.A.; Docking, C.M.; Day, J.E.L.; Breuer, K.; Edwards, S.A. Effects of species—Relevant environmental enrichment on the behaviour and productivity of finishing pigs. *Appl. Anim. Behav. Sci.* **2006**, *99*, 230–247. [CrossRef]
87. Munsterhjelm, C.; Peltoniemi, O.A.; Heinonen, M.T.; Hälli, O.; Karhapää, M.; Valros, A. Experience of moderate bedding affects behaviour of growing pigs. *Appl. Anim. Behav. Sci.* **2009**, *118*, 42–53. [CrossRef]
88. Nowicki, J.; Swierkosz, S.; Tuz, R.; Schwarz, T. The influence of aromatized environmental enrichment objects with changeable aromas on the behaviour of weaned piglets. *Vet. Arhiv.* **2015**, *85*, 425–435.
89. Sartor, K.; de Freitas, B.F.; de Souza Granja Barros, J.; Rossi, L.A. Environmental enrichment in piglet creeps: Behavior and productive performance. *BioRxiv* **2018**. [CrossRef]
90. Winfield, J.A.; Macnamara, G.F.; Macnamara, B.L.F.; Hall, E.J.S.; Ralph, C.R.; O'Shea, C.J.; Cronin, G.M. Environmental Enrichment for Sucker and Weaner Pigs: The Effect of Enrichment Block Shape on the Behavioural Interaction by Pigs with the Blocks. *Animals* **2017**, *7*, 91. [CrossRef] [PubMed]
91. Weary, D.M.; Jasper, J.; Hötzel, M.J. Understanding weaning distress. *Appl. Anim. Behav. Sci.* **2008**, *110*, 24–41. [CrossRef]
92. Di Martino, G.; Scollo, A.; Gottardo, F.; Stefani, A.L.; Schiavon, E.; Capello, K.; Marangon, S.; Bonfanti, L. The effect of tail docking on the welfare of pigs housed under challenging conditions. *Livest. Sci.* **2015**, *173*, 78–86. [CrossRef]
93. Bulens, A.; Van Beirendonck, S.; Thielen, J.; Buys, N.; Driessen, B. Long-term effects of straw blocks in pens with finishing pigs and the interaction with boar type. *Appl. Anim. Behav. Sci.* **2016**, *176*, 6–11. [CrossRef]
94. Nannoni, E.; Sardi, L.; Vitali, M.; Trevisi, E.; Ferrari, A.; Ferri, E.M.; Bacci, M.; Govoni, N.; Barbieri, S.; Martelli, G. Enrichment devices for undocked heavy pigs: Effects on animal welfare, blood parameters and production traits. *Ital. J. Anim. Sci.* **2018**, 1–12. [CrossRef]
95. Machado, S.P.; Caldara, F.R.; Foppa, L.; de Moura, R.; Gonçalves, L.M.; Garcia, R.G.; de Oliveira, G.F. Behavior of Pigs Reared in Enriched Environment: Alternatives to Extend Pigs Attention. *PLoS ONE* **2017**, *12*, e0168427. [CrossRef]

© 2019 by the authors. Licensee MDPI, Basel, Switzerland. This article is an open access article distributed under the terms and conditions of the Creative Commons Attribution (CC BY) license (http://creativecommons.org/licenses/by/4.0/).

MDPI
St. Alban-Anlage 66
4052 Basel
Switzerland
Tel. +41 61 683 77 34
Fax +41 61 302 89 18
www.mdpi.com

Animals Editorial Office
E-mail: animals@mdpi.com
www.mdpi.com/journal/animals

www.ingramcontent.com/pod-product-compliance
Lightning Source LLC
LaVergne TN
LVHW071951080526
838202LV00064B/6722